Nucleic Acid Blotting

The Basics

Series editors: R. Beynon, T. A. Brown, and C. Howe
Series advisors: T. Hunt and C. F. Higgins

DNA Sequencing
T. A. Brown

Nucleic Acid Blotting
D. C. Darling and P. M. Brickell

Somatic Cell Hybrids
C. Abbott and S. Povey

Nucleic Acid Blotting
The Basics

David C. Darling and Paul M. Brickell

The Medical Molecular Biology Unit
Department of Molecular Pathology
University College London Medical School
The Windeyer Building, Cleveland Street, London W1P 6DB

1994

Oxford University Press, Walton Street, Oxford OX2 6DP

Oxford New York Toronto
Delhi Bombay Calcutta Madras Karachi
Kuala Lumpur Singapore Hong Kong Tokyo
Nairobi Dar es Salaam Cape Town
Melbourne Auckland Madrid

and associated companies in
Berlin Ibadan

Oxford is a trade mark of Oxford University Press

Published in the United States
by Oxford University Press Inc., New York

A catalogue record for this book is available from the British Library
Library of Congress Cataloging in Publication Data
Darling, David C.
Nucleic acid blotting : the basics / David C. Darling and Paul M. Brickell. – 1st ed.
Includes bibliography references and index.
1. Nucleic acid hybridization. 2. Immunoblotting. I. Brickell, Paul M. II. Title.
QP620.D37 1994 574.87'328–dc20 94-19264
ISBN 0 19 963446 7 (Pbk)

Typeset by AMA Graphics Ltd., Preston, Lancs
Printed in Great Britain by
The Alden Press, Oxford

Foreword

Blots would be called gel transfers if it were not for the referee who rejected the paper describing the technique when it was first submitted to the *Journal of Molecular Biology* for want of significant results. Producing more data delayed publication by several months, but in the meantime, Mike Matthews took a drawing on a scrap of paper back to Cold Spring Harbor after a visit to our laboratory in Edinburgh, with a request that if they did use the method, or if they passed it on to anyone else, they acknowledge where it came from. They were scrupulous in doing this, and it was Mike who christened it 'blotting' because of its similarity to 'blotting through', the term used to describe the transfer of nucleic acid fragments from nitrocellulose strips to DEAE cellulose papers or TLC plates in Fred Sanger's 2-D fingerprinting methods that were then used for nucleic acid sequencing. Despite vigorous efforts to stamp it out, the name has stuck. The piece of paper that Mike Matthews took back with him was short on the kind of detail that you will find in this book, and, characteristically the Cold Spring Harbor group improved on the original method: applying it to full width gels rather than the single lane strips; using DNA rather than RNA for hybridization probes; and improving sensitivity to a point where single copy sequences could be detected relatively easily. It is remarkable how well the basic technique has survived, and though there have been significant changes, such as the introduction of nylon membranes, most developments have been fittingly simple—for example, using inexpensive paper towels rather than chromatography paper to soak up the transfer solution, and powdered milk instead of Denhardt's reagent to suppress background—as it is its simplicity and low cost that give the method its popularity and appeal. It is satisfying to arrive at some biologically significant result through the use of a strong solution of salt, a slab of vegetarian jelly, and a pile of paper towels.

Nevertheless, there are variants and there are wrinkles. To achieve consistently good results, accurate measurements of fragment size and quantitation of the amount of a sequence in a target population requires attention to detail and a sound knowledge of the basis of the method. The present volume is written by scientists who have a long experience of using and teaching the method and provides the experimenter with all the information needed for successful application.

E. M. Southern
Professor, Department of Biochemistry, University of Oxford

Preface

Nucleic acid blotting and hybridization techniques really are 'the basics' of molecular biology. Almost any research project that involves the study of DNA or RNA is likely to require their use at some time or other. Blotting and hybridization data frequently find their way into research papers, but these published experiments represent only the tip of a vast iceberg. Unseen are the great number of hybridization experiments performed in order to screen libraries or to check sub-cloning exercises; not to mention those performed in undergraduate classes or hospital laboratories.

Newcomers to these techniques will find instructions and help in laboratory manuals and in handbooks published by the suppliers of blotting membranes, so why write another book on the subject? For two reasons.

First, manuals are good at giving clear instructions, but rarely explain the reasoning behind each step in a procedure. Similarly, experienced workers in a lab are often happy to pass on dog-eared protocol sheets that they inherited from previous generations, but are less likely to sit down with you and explain them.

Second, manuals and protocol sheets are often rather straight-laced. They tell you the official version, but they don't tell you about the wrinkles and the short-cuts; they don't distinguish between what is essential and what is dispensable. A newcomer can usually absorb this kind of information by osmosis from others in the lab, but this takes time.

The aim of this book is therefore to help you to understand the theory behind the techniques, and so to equip you to perform them successfully and efficiently and, hopefully, to improve them. We have also indicated when there is no rational understanding of a particular step in a protocol, just the knowledge that it works. In addition, we have tried to record the folklore of the laboratories in which we have worked. This is just a part of the great oral tradition that exists in labs around the world, but we hope you will find it helpful.

When we started to write this book we planned to discuss nucleic acid electrophoresis, blotting, and hybridization. We quickly found that there was simply too much material to fit into a single volume. This book therefore concentrates on nucleic acid electrophoresis and blotting and tells you how to prepare membranes carrying DNA or RNA, ready for hybridization. A companion volume, to appear soon, will focus on hybridization and help you to complete the experiment! In the meantime, we wish you good luck.

London
July 1994

P.M.B.

Acknowledgements

We are grateful to the following colleagues in the Medical Molecular Biology Unit for allowing us to use examples of their data: Clara Ameixa (Figures 2.15, 2.17, and 2.18), Tim Brown (Figures 6.19 and 6.20), Lee Faulkner (Figure 5.6), Sarah Forbes-Robertson (Figures 2.1, 2.3, 2.4, 6.13, and 6.21), Philippa Francis (Figure 4.2), Pantelis Georgiades (Figure 4.4), David Jackson (Figure 2.2), David Latchman (Figures 1.6, 5.3 and 5.4), Torben Lund (Figures 1.4 and 2.16), Eduardo Seleiro (Figure 1.5), and John Vincent (Figure 5.5). We are grateful to Amersham International plc for providing photographs for Figures 3.3, 6.3, 6.8, and 6.16 and to Charlotte Conyers, of Amersham International, for her help. We would also like to thank Terry Brown for his encouragement and his constructive criticisms of the manuscript and the staff of Oxford University Press, for their encouragement.

Contents

How it all got started

◇ The HIV genome consists of approximately 9700 nucleotides of RNA. The haploid human genome comprises approximately 3×10^9 base pairs of DNA.

◇ A typical human gene might represent 1/200 000th of the human genome.

The genomes of even the most simple organisms are extremely complex and the analysis of complex genomes remains one of the great challenges of biology. This book and its companion address the question of how to detect a specific nucleotide sequence amongst a very complex mixture of nucleic acids; how, for example, to detect a single gene within the human genome. In this book, we will focus on the electrophoresis and blotting of DNA and RNA. In a companion volume we will discuss the hybridization of DNA and RNA blots. These techniques have taught us a great deal about the structure of complex genomes and the structure and expression of individual genes. They also have great practical value, forming the basis of a range of diagnostic methods used by geneticists, microbiologists, and pathologists.

1. The development of blotting and hybridization techniques

◇ Hydrogen bonds can be broken relatively easily.

◇ DNA can also be denatured by increasing the pH of the solution, and then renatured by lowering the pH.

◇ A 'duplex' is a double-stranded DNA molecule. A 'hybrid' is a duplex whose two strands originally came from different DNA molecules. 'Hybridization' is the process of hybrid formation.

Nucleic acid hybridization is possible because DNA molecules consist of two strands bound to each other by hydrogen bonding between complementary bases. In 1960, Marmur and Doty and their colleagues reported that heating DNA molecules in solution led to the dissociation of the two strands (denaturation) and that subsequent cooling under controlled conditions led to their reassociation (renaturation), as shown in *Figure 1.1* (Marmur and Lane 1960; Doty *et al.* 1960). It was soon realized, by them and others, that DNA strands from different sources could reassociate in solution to form hybrid duplexes, as long as the two strands had complementary nucleotide sequences (reviewed by McCarthy and Church 1970). It

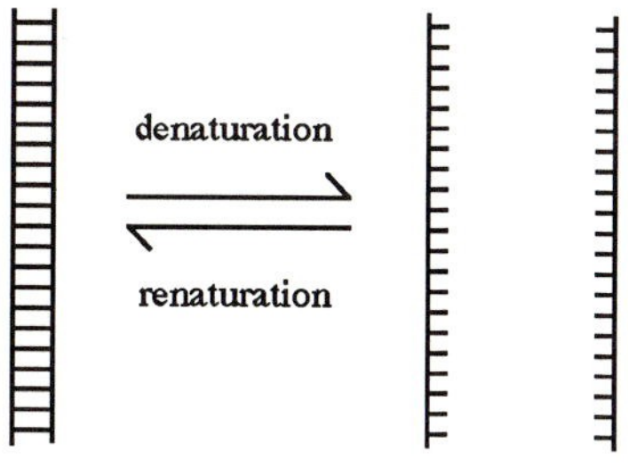

Fig 1.1

Denaturation and renaturation of DNA. DNA can be denatured by increasing the temperature of the solution above 90 °C or by raising its pH above 13. If the solution is cooled, or its pH returned to neutral, the DNA will renature

was also found that DNA strands could form hybrids with complementary RNA strands. The 1960s saw a rapid growth in the use of these 'solution hybridization' techniques to explore genome structure and expression. Such experiments led, for example, to the discovery of highly repeated sequences in eukaryotic genomes (Britten and Kohne 1968) and of different abundance classes of mRNA in eukaryotic cells (Bishop *et al.* 1974).

One of the problems encountered in performing DNA–RNA hybridization experiments in solution was that reassociation of the two DNA strands competed with the formation of DNA–RNA hybrids. One way to overcome this was to bind the denatured DNA strands to an insoluble support so that they were unable to reassociate with each other but could hybridize to complementary RNA strands in the surrounding solution. Since DNA was known to bind efficiently to pieces of nitrocellulose filter, these became widely used, giving birth to the 'filter hybridization' techniques that are such a central component of modern molecular biology.

1.1 Southern blotting

By the early 1970s, it had been found that restriction endonucleases could be used to cut large DNA molecules into smaller fragments and that agarose or polyacrylamide gel electrophoresis could be used to separate these fragments according to their size. Among the investigators using this new technology were those who wanted to determine the structure and organization of the DNA sequences that are transcribed into RNA. This meant that they had to identify specific DNA fragments that hybridized to particular RNA species. A typical experiment (*Figure 1.2*) would be to digest DNA with a restriction endonuclease, electrophorese the fragments on an agarose gel and then slice the gel into pieces, each containing a narrow size-range of DNA fragments. The DNA fragments were then eluted from each gel slice and hybridized to radiolabelled RNA, either in solution or after binding the DNA fragments to nitrocellulose. The amount of radioactivity binding to the DNA fragments from each gel slice was then counted and a picture of the gel reconstructed to show where the peak(s) of radioactivity lay. This gave a rough estimate of the size of the DNA fragment(s) to which the radiolabelled RNA bound.

◇ This sounds like hard work, and it is.

The limits of the technology, and remember that DNA cloning was still waiting in the wings, meant that only very abundant RNA species, such as certain viral mRNAs and cellular ribosomal RNAs, could be analysed in this way. The poor resolution and the appalling amount of work involved in analysing all of the gel slices, provided a strong impetus for the development of improved technology. This came in 1975 when Ed Southern described the blotting technique that now bears his name. In this technique (*Figure 1.4*), eukaryotic genomic DNA was digested with a restriction endonuclease. The fragments were then electrophoresed in an agarose gel, and transferred directly on to a strip of nitrocellulose filter. The bound DNA

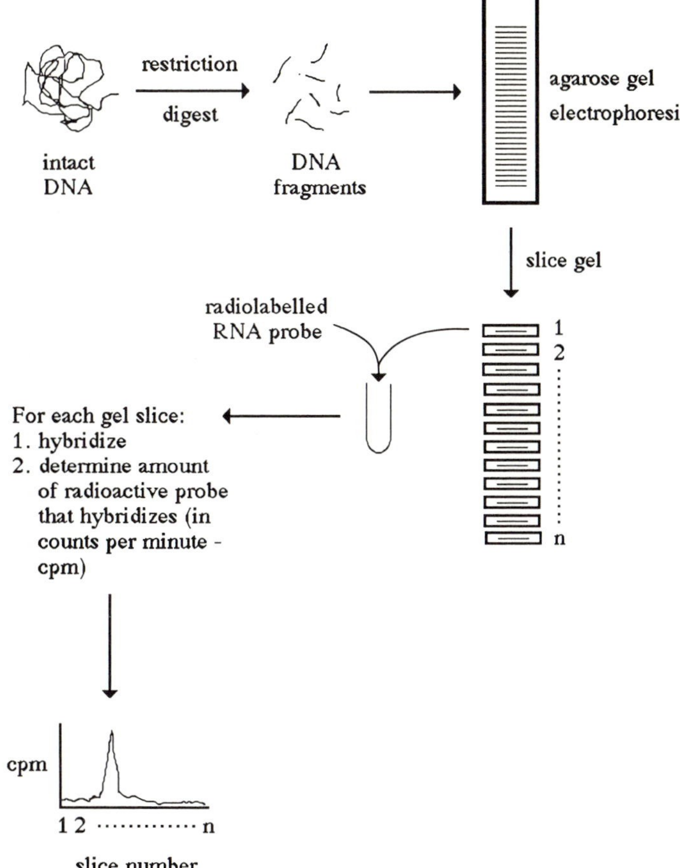

Fig 1.2

Hybridization to electrophoresed nucleic acids in the 'bad old days', before Southern blotting was invented

◇ Nitrocellulose sheets were developed for filtration and borrowed for blotting. Even nitrocellulose sheets manufactured specifically for blotting are usually called *filters*. Nylon sheets manufactured for blotting are usually called *membranes*.

◇ Fluorography is used to enhance autoradiographic signals produced by weak β-emitters, such as ^{3}H, ^{14}C, and ^{35}S. The filter is impregnated with a chemical (a fluor) that emits light when struck by β-radiation. This light exposes the film.

◇ In southern blotting, DNA is transferred from a gel to a nitrocellulose filter or nylon membrane.

was then hybridized with ^{3}H-labelled 28S ribosomal RNA. This had been labelled *in vivo*, by growing cells in the presence of ^{3}H-labelled precursors, and then isolated from the cells by physical means. DNA fragments that bound labelled RNA were visualized by fluorography of the nitrocellulose filter. With the rapid expansion in the use of DNA cloning techniques that was about to take place, Southern blotting became very widely used. One of the early successes of the technique, and perhaps its most dramatic, was the discovery of introns by Jeffreys and Flavell in 1977. *Figure 1.3* shows the result of a typical Southern blot.

1.2 Northern blotting

Meanwhile, others were perfecting methods for blotting RNA on to filters. Nitrocellulose filters bind RNA rather poorly, but it was found that RNA would bind covalently to finely divided cellulose particles to which diazobenzyloxymethyl (DBM) groups had been attached ('derivatization'). A technique was therefore devised for transferring RNA to cellulose filter paper sheets that had been derivatized with DBM (Alwine *et al.* 1977). This rapidly became known

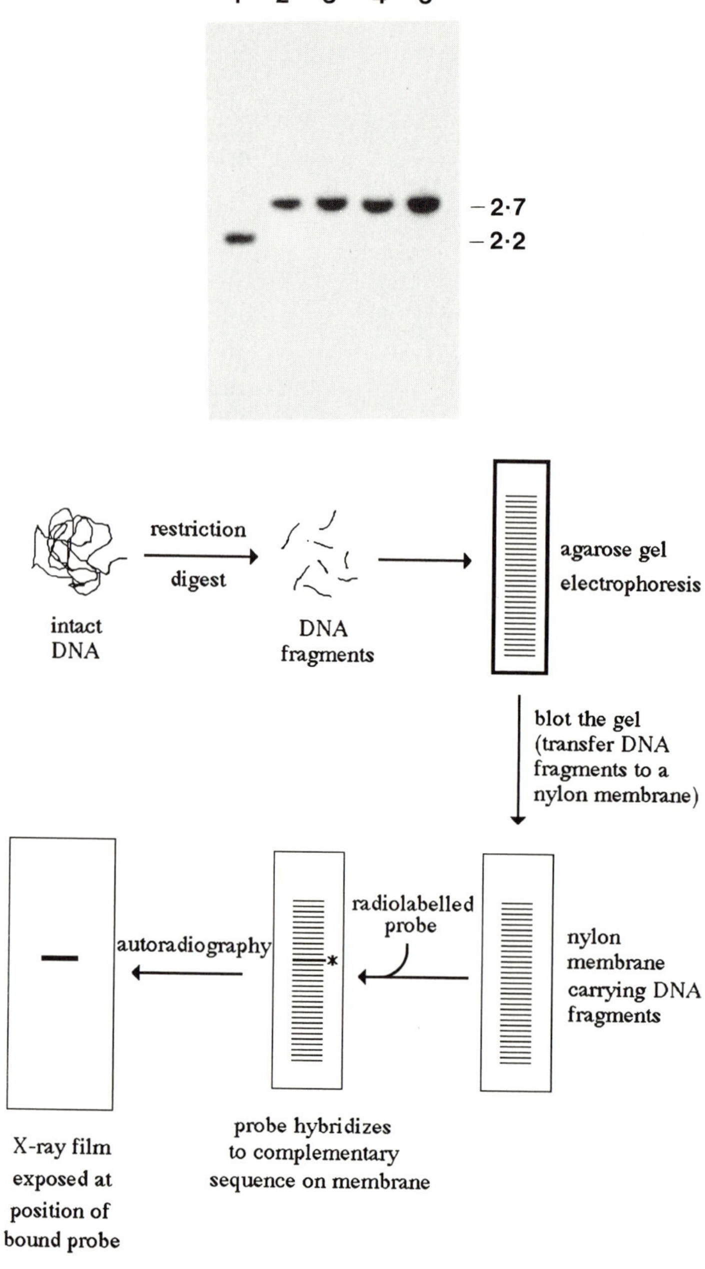

Fig 1.3

A Southern blot. Genomic DNA from five mouse strains (1–5) was digested with *PstI*, electrophoresed on an agarose gel, blotted on to a nylon and hybridized with a ^{32}P-labelled probe for the H-2Eα gene. The probe detects a restriction fragment length polymorphism. Sizes of DNA fragments are shown in kilobase pairs. (Lund *et al.* 1990)

Fig 1.4

Southern blotting

◇ In northern blotting, RNA is transferred from a gel to a nitrocellulose filter or nylon membrane.

as northern blotting, a joke that nobody seems prepared to own up to. *Figure 1.5* shows the result of a typical northern blot.

1.3 Further advances

For the full potential of these blotting techniques to be realized, a number of other developments had to take place. Most important, was the development of a method for *in vitro* synthesis of radiola-

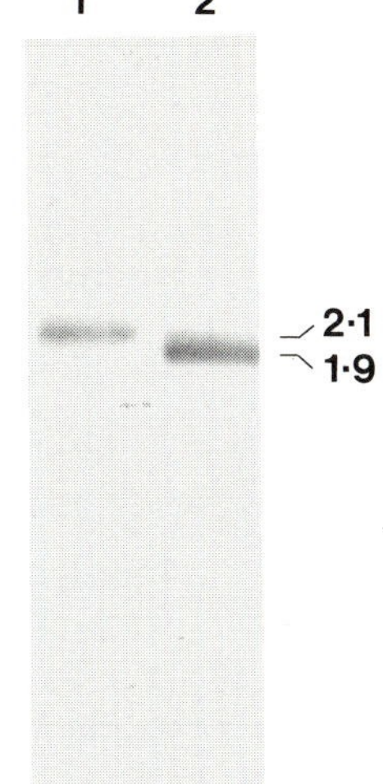

Fig 1.5

A northern blot. Polyadenylated RNA from chick embryo eye (1) or liver (2) was electrophoresed on an agarose gel, blotted on to nylon and hybridized with a ^{32}P-labelled probe for retinoid X receptor-γmRNA. The probe detects an mRNA of 2.1 kilobases in the eye, and a range of mRNAs (1.9–2.1 kilobases) in the liver.

◇ This classic paper starts with a description of how to make your own ^{32}P-labelled deoxynucleoside triphosphates, starting with deoxynucleosides and 50 mCi of ^{32}P-labelled H_3PO_4. Thankfully, those days have gone.

belled nucleic acid probes that had high specific activities. The technique of 'nick translation', described by Rigby *et al.* (1977), became the standard means of labelling double-stranded DNA with ^{32}P and held absolute sway until the development of the 'random-priming' method by Feinberg and Vogelstein in 1984. In the meantime, methods were also developed for labelling single-stranded DNA and RNA to high specific activity. Also important was the refinement of highly sensitive autoradiographic techniques for the detection of ^{32}P and other isotopes (Laskey and Mills 1977).

Filter immobilization was also adapted for the detection of recombinant plasmids in bacterial colonies (Grunstein and Hogness 1975) and, later, of recombinant bacteriophages in viral plaques (Benton and Davis 1977). A related technique for immobilizing DNA and RNA on filters without electrophoresing them beforehand became known as 'dot blotting', and a later modification of this was called 'slot blotting'.

◇ Western blotting is a technique for transferring proteins from a gel to a nitrocellulose filter.

More recently, the 'western blotting' technique has been developed for transferring proteins from SDS-polyacrylamide gels to nitrocellulose filters so that they can be detected with labelled antibodies or other protein-binding molecules (Towbin *et al.* 1979). *Figure 1.6* shows the result of a typical western blot. DNA sequences that bind nuclear proteins have been identified by transferring double-stranded DNA molecules from agarose gels to filters and incubating them with labelled proteins extracted from cell nuclei. This procedure is called 'south-western blotting'. A method for blotting polyadenylated mRNA from agarose gels on to cellulose filter paper coupled with polyuridylic acid, which was developed at Tel Aviv University, enjoyed brief fame as 'Middle-Eastern blotting' (Wreschner and Herzberg 1984). These techniques will not be discussed in this book but are mentioned to illustrate how much duller the world would be had the Southern blot been invented by somebody named Smith.

The spread of blotting techniques stimulated the development and refinement of new solid supports to which DNA and RNA could be

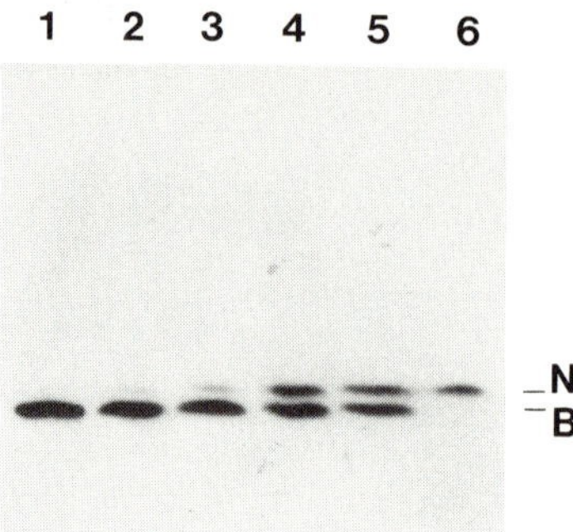

Fig 1.6

A western blot. Protein was isolated from rat brain at embryonic days 16 (1), 17 (2), and 18 (3), at days 2 (4) and 7 (5) post-partum and in adulthood (6). Protein was electrophoresed on an SDS-polyacrylamide gel, blotted on to nitrocellulose and incubated with antiserum against two related proteins, SmN and SmB. (Grimaldi *et al.* 1993)

transferred. In particular, nylon membranes were introduced to solve some of the problems associated with the use of nitrocellulose filters. A wide range of nitrocellulose filters and nylon membranes are now available.

2. This book

In the next five chapters, we will describe the range of techniques for immobilizing nucleic acids on membranes. We will discuss DNA electrophoresis (Chapter 2) and Southern blotting (Chapter 3), RNA electrophoresis and northern blotting (Chapter 4), dot and slot blotting (Chapter 5) and Benton-and-Davis and Grunstein–Hogness screening of bacterial colonies and bacteriophage plaques, respectively (Chapter 6). The types of nitrocellulose filters and nylon membranes that are available, and the factors that determine which should be used in a particular situation, will be discussed in Chapter 7. Once membranes have been prepared by one of the methods described in Chapters 2 to 6, the procedures for detecting immobilized sequences by nucleic acid hybridization are very similar. The preparation of nucleic acid probes and the hybridization procedure itself will be discussed in a companion volume in this series.

3. Further reading

These are general laboratory manuals that include detailed protocols for the techniques to be discussed in this book. The relevant section of each manual will be cited at the end of each chapter of this book.

◇ Most manufacturers of nylon membranes supply instruction manuals with their products.

Berger, S.L. and Kimmel, A.R. (ed.) (1987). Guide to molecular cloning techniques. *Methods in Enzymology*, **152**.

Perbal, B. (1988). *A practical guide to molecular cloning* (2nd edn). Wiley, New York.

Sambrook, J., Fritsch, E.F., and Maniatis, T. (1989). *Molecular cloning: a laboratory manual* (2nd edn). Cold Spring Harbor Laboratory Press.

◇ This manual, in three volumes, is the most comprehensive of them all. It is universally known as 'Maniatis', because the first edition was by Maniatis *et al.*

4. Laboratory safety

◇ You must read and follow your laboratory's codes of safe practice.

Perbal, B. (1988). *A practical guide to molecular cloning*, (2nd edn), pp. 4–10. Wiley, New York.

Zoon, R.A. (1987). Safety with ^{32}P- and ^{35}S-labeled compounds. *Methods in Enzymology*, **152**, 25–29.

5. References

Alwine, J.C., Kemp, D.J., and Stark, G.R. (1977). Method for detection of specific RNAs in agarose gels by transfer to diazobenzyloxymethyl-paper and hybridization with DNA probes. *Proceedings of the National Academy of Sciences, USA*, **74**, 5350–54.

Benton, W.D. and Davis, R.W. (1977). Screening λgt recombinant clones by hybridization to single plaques in situ. *Science*, **196**, 180–2.

Bishop, J.O., Morton, J.G., Rosbash, M., and Richardson, M. (1974). Three abundance classes of HeLa cell messenger RNA. *Nature*, **250**, 199–204.

Britten, R.J. and Kohne, E.D. (1968). Repeated sequences in DNA. *Science*, **161**, 529–40.

Doty, P., Marmur, J., Eigner, J. and Schildkraut, C. (1960). Strand separation and specific recombination in deoxyribonucleic acids: physical chemical studies. *Proceedings of the National Academy of Sciences, USA*, **46**, 461–76.

Feinberg, A.P. and Vogelstein, B. (1984). A technique for radiolabeling DNA restriction endonuclease fragments to high specific activity. Analytical Biochemistry, **137**, 266–7.

Grimaldi, K., Horn, D.A., Hudson, L.D., Terenghi, G., Barton, P., Polak, J.M., and Latchman, D.S. (1993). Expression of the SmN splicing protein developmentally regulated in the rodent brain but not in the rodent heart. *Developmental Biology*, **156**, 319–23.

Grunstein, M. and Hogness, D.S. (1975). Colony hybridization: a method for the isolation of cloned DNAs that contain a specific gene. *Proceedings of the National Academy of Sciences, USA*, **72**, 3961–5.

Jeffreys, A.J. and Flavell, R.A. (1977). The rabbit β-globin gene contains a large insert in the coding sequence. *Cell*, **12**, 1097–1108.

Laskey, R.A. and Mills, A.D. (1977). Enhanced autoradiographic detection of ^{32}P and ^{125}I using intensifying screens and hypersensitized film. FEBS Letters, **82**, 314–16.

Lund, T., Simpson, E., and Cooke, A. (1990). Restriction fragment length polymorphisms in the major histocompatibility complex of the non-obese diabetic mouse. *Journal of Autoimmunity*, **3**, 289–98.

Marmur, J. and Lane, D. (1960). Strand separation and specific recombination in deoxyribonucleic acids: biological studies. *Proceedings of the National Academy of Sciences, USA*, **46**, 453–61.

McCarthy, B.J. and Church, R.B. (1970). The specificity of molecular hybridization reactions. *Annual Reviews of Biochemistry*, **39**, 131–50.

Rigby, P.W.J., Dieckmann, M., Rhodes, C., and Berg, P. (1977). Labeling deoxyribonucleic acid to high specific activity *in vitro* by nick translation with DNA polymerase I. *Journal of Molecular Biology*, **113**, 237–51.

Southern, E.M. (1975). Detection of specific sequences among DNA fragments separated by gel electrophoresis. *Journal of Molecular Biology*, **98**, 503–17.

Towbin, H., Staehelin, T., and Gordon, J. (1979). Electrophoretic transfer of proteins from polyacrylamide gels to nitrocellulose sheets: procedure and some applications. *Proceedings of the National Academy of Science, USA*, **76**, 4350–4.

Wreschner, D.H. and Herzberg, M. (1984). A new blotting medium for the simple isolation and identification of highly resolved messenger RNA. *Nucleic Acids Research*, **12**, 1349–59.

2 Southern blotting I: electrophoresis of DNA

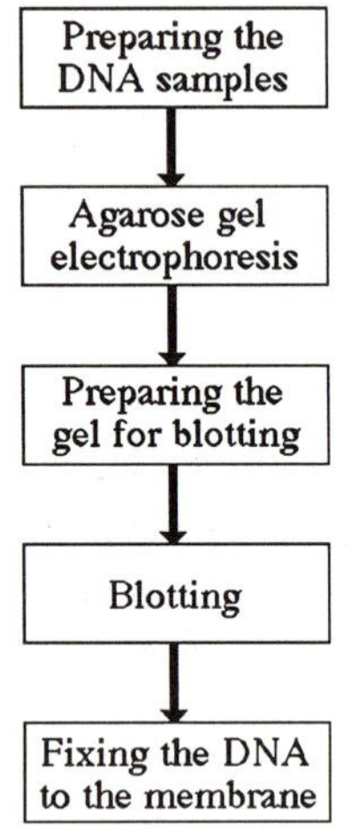

Southern blotting (Southern 1975) is a procedure developed for transferring single-stranded DNA and denatured double-stranded DNA from an agarose gel to a nylon membrane or nitrocellulose filter. The blotted DNA is tightly bound to the membrane but can still hybridize to a labelled nucleic acid probe. The technique can be adapted to transfer DNA from a polyacrylamide gel (Smith *et al.* 1984), but we will not discuss this in detail. The steps involved in Southern blotting from an agarose gel are:

(1) preparation of the DNA samples for electrophoresis,

(2) agarose gel electrophoresis,

(3) preparation of the agarose gel for blotting,

(4) blotting,

(5) fixing the DNA to the membrane.

◇ Agarose is a linear polysaccharide extracted from seaweed.

Agarose gels used for Southern blotting consist of horizontal slabs of agarose. They are made by dissolving agarose powder by boiling in an appropriate buffer solution, pouring the molten gel into a plastic gel-casting tray and allowing it to set. A plastic comb is used to create slots, or wells, in the gel to receive the DNA samples. When set, the gel consists of a mesh of agarose molecules, with a pore size that depends on the agarose concentration. The gel is submerged in running buffer and the DNA samples are loaded into the wells. An electrical field is then applied across the gel and the negatively charged DNA molecules migrate towards the positive terminal. The rate at which linear DNA molecules migrate in a gel depends on their length. Larger molecules migrate more slowly than smaller molecules because it is more difficult for them to pass through the pores. The DNA molecules therefore become separated in the gel, according to their size. DNA is visualized by including ethidium bromide in the gel or by soaking the gel in a solution of ethidium bromide after electrophoresis. Ethidium bromide is a fluorescent dye that intercalates between the two strands of DNA and makes them visible under ultraviolet (UV) radiation (*Figure 2.1*).

◇ The rate of migration of a linear DNA molecule is inversely proportional to the $\log_{10}$ of its length in base pairs.

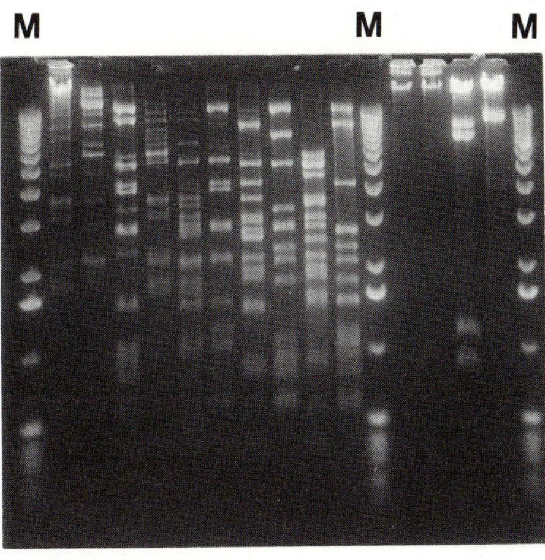

Fig 2.1

Samples of a cosmid genomic clone were digested with restriction endonucleases and electrophoresed on an agarose gel containing ethidium bromide. When placed on a transilluminator, DNA fragments are visible as bright bands. Tracks labelled M contain DNA size markers ('1 kb DNA ladder' from Gibco-BRL, catalogue number 520-5615SA)

The way in which you carry out the agarose gel electrophoresis step will be determined to some extent by the source of your DNA samples and so we will first discuss the different types of DNA that you are most likely to use and how they should be handled.

1. Different sources of DNA

We will assume that you have already isolated good quality DNA. DNA can come from a wide range of sources and in many different forms. You are most likely to be using one of the following:

◇ 1 bp = 1 base pair
1 kb = 1 kilobase pair = 1000 base pairs
1 Mb = 1 megabase pair = 1000 kb

- Genomic DNA—the primary genetic material in most organisms. When uncut, your extract of genomic DNA may contain extremely large molecules, typically greater than 1000 kb in length (Birnboim 1992; Gross-Bellard *et al.* 1973; Guidet and Langridge 1992). When digested with restriction enzymes, DNA fragments may range from 10–20 bp to greater than 1000 kb.
- Plasmid DNA—small circular DNA molecules, propagated in bacteria. These can be quite large (up to 200 kb), but you are most likely to be using plasmids ranging from 2 to 10 kb.
- Bacteriophage λ DNA—linear DNA molecules of approximately 50 kb, propagated as viruses that infect bacteria.
- Cosmid DNA—circular plasmids containing the terminal *cos* sequence of bacteriophage λ (Ish-Horowitz and Burke 1981). Recombinant cosmids are approximately 50 kb long.
- Yeast artificial chromosome (YAC) DNA—plasmids that can be propagated as chromosomes in yeast (Burke *et al.* 1987). Recombinant YACs are typically several hundred kb in length and may exceed 1 Mb.
- PCR product DNA—DNA fragments resulting from amplification by the polymerase chain reaction. These can be up to approximately 3 kb long.

◇ Plasmid vectors are typically used to construct cDNA libraries and for cloning fragments of cDNA and genomic DNA.

◇ Bacteriophage λ vectors are used to construct cDNA and genomic libraries.

◇ Cosmid vectors are used to construct genomic libraries.

◇ YAC vectors are used to construct genomic libraries, one of their advantages being that a relatively small number of YAC clones are needed for a 'complete' library.

The size of the DNA that you are using can therefore be as small as 20 bp or as large as several hundred kb. Different strategies are required for the electrophoresis of DNA molecules of such varying sizes. In particular, you will need to decide how much DNA to load on to the gel and what concentration of agarose to use in the gel.

1.1 Genomic DNA

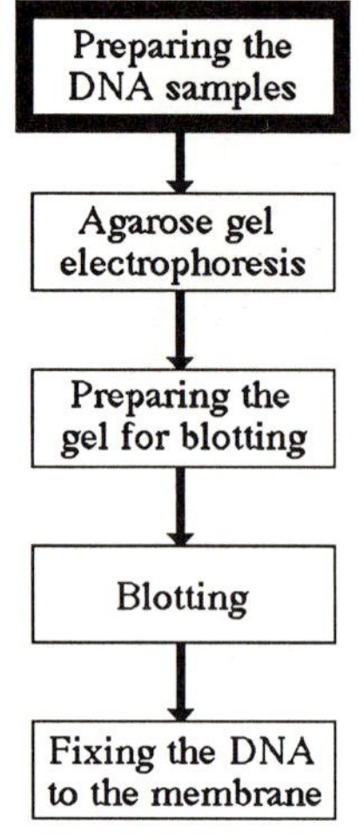

You must first digest an appropriate amount of genomic DNA with a restriction enzyme (Bhagwat 1992). We will not discuss general aspects of performing restriction digests, but only those features relevant to subsequent agarose gel electrophoresis.

The amount of DNA you should digest depends on the objective of your experiment. If you want to use Southern blotting to detect a single-copy sequence in a mammalian genome, using a standard radiolabelled probe that is several hundred base pairs long, you should load 10–20 μg of digested DNA in a standard gel well (typically measuring 3 × 1 mm). We find that you get sharper bands if you use wider wells, for example by taping two teeth of the well-forming comb together. In this case, load proportionately more DNA per well. If you plan to use a very short probe, such as a labelled oligonucleotide, you must load more DNA, because the sensitivity of the method will be lower than with a long probe. For experiments of this sort, load 30–50 μg of digested DNA in a 3 × 1 mm well.

◇ If you load too much DNA you will get poor quality bands.

It is wise to check that your restriction digest has worked properly before running the gel for Southern blotting. To do this, take a small aliquot containing approximately 0.5 μg of DNA, add gel loading buffer and electrophorese the sample on a small (for example, 10 × 10 cm) agarose gel, alongside a sample of undigested DNA. The features to look for in the electrophoresed sample are discussed in section 3.9.1.

Once you have checked that your genomic DNA is fully digested, simply add loading buffer to the remainder of the digestion mixture and load all of the sample, as described in section 3.5, on to the gel that will be blotted. One problem that you might encounter is that since genomic DNA is often best digested at a fairly low concentration, the volume of your digest may be too large to fit into the well. If this is the case, concentrate the digested DNA by precipitating it in ethanol and resuspending it in a suitably small volume of loading buffer. Make sure that there is no residual ethanol in the DNA solution. If there is, the DNA will float out of the well during loading, causing you great distress. To be sure that all of the ethanol has gone, dry the pellet of precipitated DNA thoroughly, before resuspension.

◇ Some people also incubate the resuspended DNA at 70 °C for 10 minutes before loading.

1.2 Plasmid DNA

The amount of plasmid DNA you should load depends on the size of the insert of interest. If, for example, your recombinant plasmid comprises 3 kb of vector and 1 kb of insert, then 1 μg of digested

◇ With a genomic Southern blot, the DNA fragment to which the probe hybridizes is present in limited quantities. With plasmid Southern blots this is not so and you can get good results using low specific activity probes and short autoradiography times.

plasmid DNA will yield only 250 ng of insert. Similarly, if the insert is only 250 bp long then 1 μg of the digested plasmid will yield only 77 ng of insert. In practice, it is usually most convenient to load 1–2 μg of digested plasmid DNA in a well. Inserts of all sizes should then be easily detectable by Southern blotting, even though very small inserts may be difficult to see on the ethidium bromide-stained gel.

1.3 Bacteriophage λ and cosmid DNA

Clones constructed in bacteriophage λ replacement vectors or cosmid vectors present special problems because of the large sizes of their inserts. When these vectors are digested with restriction enzymes the insert DNA will probably yield multiple fragments, ranging in size from 10 to 20 bp up to approximately 20 kb, in the case of bacteriophage λ vectors (*Figure 2.2*), or approximately 40 kb, in the case of cosmid vectors (*Figure 2.1*). It is often not possible to separate all of these fragments on a single gel. In addition, it is easy to run potentially interesting fragments off the end of the gel as they may be too small to be visible (see *Figure 2.2*). One option is to electrophorese part of the digested sample for a long period, in order to separate the larger fragments properly, and to electrophorese the rest of the sample for a shorter period, in order to preserve the smaller fragments. This can be done on the same gel by staggering the time at which the samples are loaded. Remember that if the fragment of interest is 1 kb long, and the vector and insert together are 50 kb long, then when 1 μg of digested DNA is loaded, only 20 ng of the fragment of interest will be present. This will be readily detectable by Southern blotting and hybridization, but will not be visible on the ethidium bromide-stained gel. A similar situation is encountered when using bacteriophage λ insertion vectors, in which the insert will represent only a small part of the whole recombinant DNA molecule.

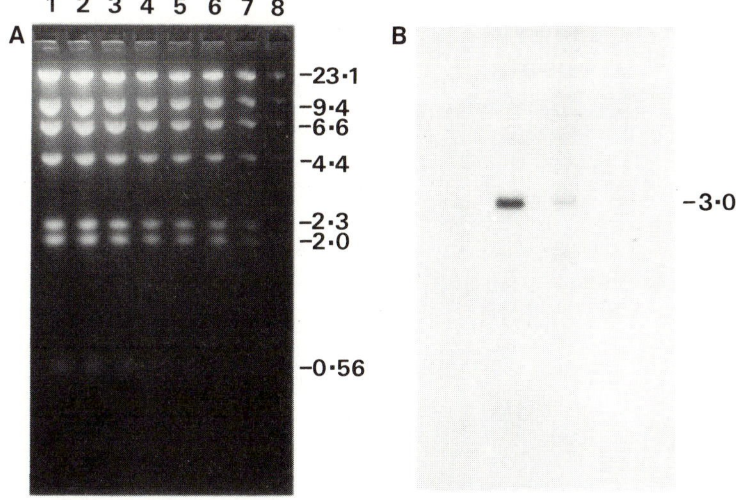

Fig 2.2

A. Different amounts of *Hind*III-digested bacterophage λ DNA were electrophoresed in an agarose gel. Sizes of DNA fragments are shown in kb. A 125 bp fragment is present in all tracks, but is not visible. The 560 bp fragment is present, but not visible, in tracks 6, 7, and 8. The 2 and 2.3 kb fragments are present, but not visible, in track 8. Larger fragments are overloaded in tracks 1, 2, and 3. As well as λ DNA, tracks 3 and 4 contain 1 ng and 100 pg of linearized plasmid DNA, respectively. The plasmid is 3 kb long but is not visible in either track. B. The gel in A was blotted on to nylon and probed with radiolabelled plasmid DNA, which detected the plasmid DNA in tracks 3 and 4, even though it was not visible on the gel

1.4 YAC DNA

YAC DNA can be huge; so long in fact that special types of electrophoresis are required to analyse it. New approaches to DNA technology, such as pulsed-field gel electrophoresis (PFGE; Smith and Cantor 1986), are evolving to permit analysis of these and other very large DNA molecules. Such molecules can be blotted and hybridized, using modified versions of the standard techniques. We will not discuss these here.

1.5 PCR product DNA

◇ Unused oligonucleotide primers will run at the bottom of the gel.

Amplified DNA products from PCR can be up to 3 kb in length but are rarely larger, because of the limitations of PCR. PCR may give rise to a single DNA fragment, to multiple fragments or to a smear of fragments, depending on the type of PCR performed. These products require no further treatment prior to loading on to the gel.

2. Before you load the DNA on to the gel

Once the correct amount of DNA has been digested, add loading buffer. This is usually made up at 6 × or 10 × the final concentration. Thus, if you have digested the DNA in a volume of 10 μl you should add 2 μl of 6 × buffer or 1.1 μl of 10 × buffer.

◇ There are many different types of loading buffer, but they all do the same job.

Loading buffers contain one or more dyes to help you see that the whole sample is loaded into the well without spillage, and to allow you to see at a glance the progress of electrophoresis. It is most common to use bromophenol blue, xylene cyanol FF or orange G, alone or in combination. Like DNA fragments, these dyes migrate from the negative terminal to the positive terminal during electrophoresis, and do so at characteristic rates. The length of DNA fragment with which a dye co-migrates can vary with buffer conditions, agarose concentration and even with the brand of agarose used. In *Figure 2.3*, xylene cyanol FF and bromophenol blue are co-migrating with DNA fragments of approximately 4000 and 600 base pairs, respectively. Loading buffers also contain sucrose, glycerol or Ficoll in order to make the sample denser than water, so that it will sink to the bottom of the well when loaded on to a gel submerged under running buffer. This makes loading much easier.

◇ Always use a fresh micropipette tip whenever you remove an aliquot of loading buffer from the stock.

Keep a stock of loading buffer, rather than making a fresh batch each time, but be very careful not to contaminate this stock with DNA. In particular, traces of contaminating plasmid DNA that are invisible on the gel may be detectable following Southern blotting and hybridization with a probe that contains plasmid sequences. This is not an uncommon source of unexpected extra bands on autoradiographs of hybridized blots (*Figure 2.2*).

When you have added loading buffer to the digested DNA sample, mix them together well. Do this by gentle mixing with a micro-

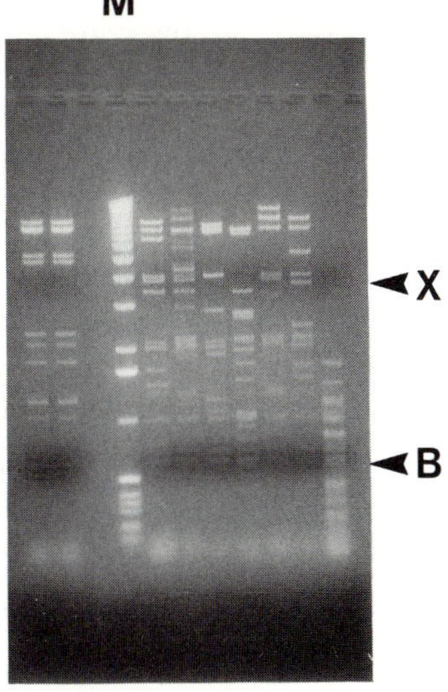

Fig 2.3

A 0.8% agarose gel on which DNA has been electrophoresed. The photograph is over-exposed, to show the xylene cyanol FF (X) and bromophenol blue (B) dyes from the loading buffer. Comparison with DNA size markers (track M) shows that on this gel, xylene cyanol FF and bromophenol blue co-migrate with DNA fragments of approximately 4 kb and 600 bp, respectively

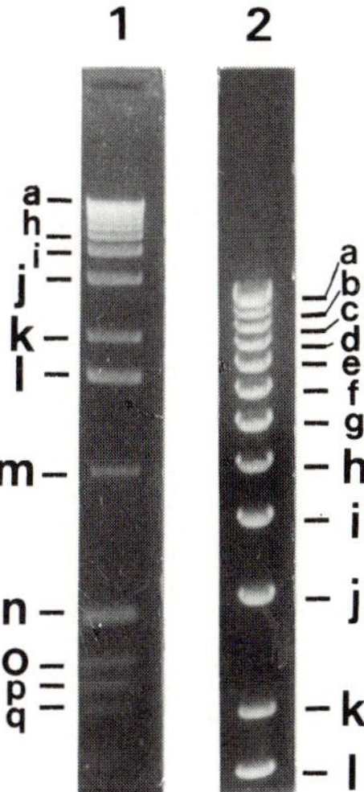

Fig 2.4

Tracks 1 and 2 contain '1 kb DNA ladder' (Gibco-BRL, catalogue number 520-5615SA), which contains 23 DNA fragments. Their sizes (in bp) are: (a), 12216; (b), 11198; (c), 10180; (d), 9162; (e), 8144; (f), 7126; (g), 6108; (h), 5090; (i), 4072; (j), 3054; (k), 2036; (l), 1636; (m), 1108; (n), 517; (o), 506; (p), 396; (q), 344; (r), 298; (s), 220; (t), 201; (u), 154; (v), 134; (w), 75. In track 1, fragments r–w are invisible and fragments a–g have not separated. The sample in track 2 has been electrophoresed for longer. Fragments m–w have run off the end of the gel, but the remaining fragments, including a–g, can now be resolved

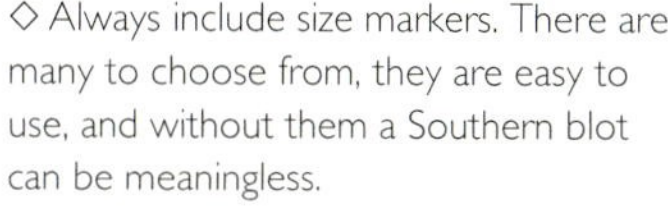
◇ Always include size markers. There are many to choose from, they are easy to use, and without them a Southern blot can be meaningless.

pipette or by gentle vortexing. Any air bubbles generated must be removed by spinning for a minute in a microcentrifuge.

As well as preparing DNA samples, prepare size markers so that you can later estimate the sizes of the DNA fragments in your samples. The best size markers to use will depend on the size of the DNA fragment of interest. A favourite of many workers is *Hind*III-digested bacteriophage λ DNA. This contains DNA fragments that range in size from 125 bp to 23.1 kb and is suitable for most purposes (*Figure 2.2*). There are also a number of commercially available mixtures of DNA fragments that have been developed specifically for use as size markers. These include the '1 kb DNA ladder', marketed by Gibco-BRL (*Figures 2.1* and *2.4*). Load sufficient size marker DNA to be visible on the ethidium bromide-stained gel. Between 05 and 1 μg of either of the above markers is usually adequate. Remem-

◇ It may be worthwhile running different amounts of your size markers in a couple of different tracks.

ber that with mixtures of fragments of widely differing sizes, such as these, it is inevitable that if one particular fragment is present at an optimal level, then others will be either overloaded or underloaded. When fragments are overloaded they give broad bands that run faster than they should, introducing inaccuracies into your size determinations (*Figure 2.2*). When fragments are underloaded, they are invisible in the stained gel (*Figure 2.2*).

Some people load tiny amounts of size markers on their gel and then 'spike' their hybridization probe with radiolabelled size marker. The positions of the size marker fragments are then visible on the autoradiograph. In principle, this should make it easier to compare the sizes of the hybridizing fragments in your sample tracks with those of the marker fragments. In practice, we have never found this to be very satisfactory. The different size marker fragments often hybridize to greatly differing extents, with some being invisible, and extra bands sometimes appear. This makes interpretation very difficult. We recommend that you rely on size markers that are clearly visible on the ethidium bromide-stained gel.

3. Preparing and running the gel

3.1 Agarose

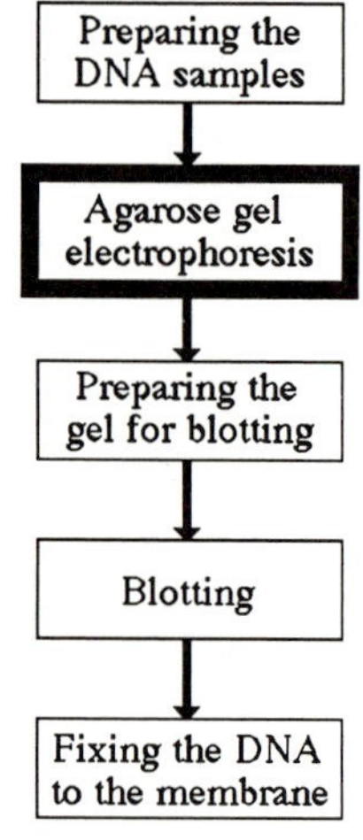

Use good quality, molecular biology grade agarose for making gels because this is free from contaminants that could interfere with electrophoresis or degrade the electrophoresed DNA. Do *not* use low melting point (LMP) agarose because gels made with this are fragile and will fall apart during the manipulations involved in Southern blotting.

The rate at which DNA molecules migrate depends on the concentration of agarose in the gel and so you must decide what concentration to use. The highest concentration of agarose that it is feasible to use is about 2 per cent (w/v) and the lowest is about 0.5 per cent (w/v). High percentage gels tend to compress the bands formed by higher molecular weight DNA fragments and improve the resolution of smaller fragments. Conversely, lower percentage gels allow better separation of the larger fragments, whilst tending to compress the smaller fragments. Thus, by using gels with different agarose concentrations it is possible to resolve DNA fragments with a wide range of sizes. *Table 2.1* shows the size ranges of DNA fragments that are separated optimally by different concentrations of agarose. For most purposes 0.8 per cent (w/v) or 1 per cent (w/v) agarose gels are used. These give good resolution of DNA fragments over a range of sizes that is appropriate for restricted DNA, are relatively robust and easy to handle, and allow efficient Southern blotting of DNA on to membranes.

◇ At low agarose concentrations, gels become difficult to handle and break easily.

Table 2.1 Size ranges of linear DNA fragments that are separated optimally by different concentrations of agarose (data from Perbal 1988)

Concentration of agarose (% w/v)	Range of sizes that are separated optimally (kb)
0.3	5.0–60
0.6	1.0–20
0.8	0.8–10
1.0	0.4–8
1.2	0.3–7
1.5	0.2–4
2.0	0.1–3

3.2 Electrophoresis buffers

Electrophoresis is performed at, or close to, neutral pH so that the nucleic acid molecules remain negatively charged and double stranded. The gel must therefore be poured and electrophoresed in a solution with adequate buffering capacity. Electrophoresis is also affected by the ionic strength of the buffer. If you forget to put buffer in the gel (and we have), then very little current will flow and the DNA will migrate extremely slowly. Conversely, if the ionic strength is too high, perhaps because you used concentrated stock buffer solution rather than diluted working strength solution (and we have done that too), the current will be very high and the gel will heat up and may begin to melt.

◇ 1 × TAE buffer consists of 0.04 M tris-acetate, pH 7.6, 0.001 M EDTA. 1 × TBE buffer consists of 0.089 M Tris-borate, pH 8.3, 0.0025 M disodium EDTA. EDTA chelates divalent cations and so inactivates divalent cation-dependent deoxyribonucleases (DNases), which degrade DNA.

There are a number of different electrophoresis buffers that you could use. The most commonly used are **TBE** (Tris-borate-EDTA) and **TAE** (Tris-acetate-EDTA). Both buffers maintain a pH of between 7.5 and 7.8, although TAE has a lower buffering capacity than TBE and is more easily exhausted during long periods of electrophoresis. Double-stranded linear DNA molecules migrate a little faster through TAE gels than through TBE gels, but the abilities of gels run in the two buffers to resolve linear DNA molecules are very similar. It has been claimed that Southern blotting is more efficient from TAE gels than from TBE gels, but we have not encountered any problems in blotting from TBE gels. It is therefore largely a matter of personal choice whether you use TBE or TAE for electrophoresis prior to Southern blotting, and both buffers have their adherents in our laboratory.

◇ Store TAE and TBE stocks at room temperature and dilute them before use.

TBE is usually made up as a 10 × concentrated stock solution. A precipitate tends to form in TBE stocks over a period of time and the stock should be discarded if this becomes too heavy. TAE is usually kept as a 50 × concentrated stock.

3.3 Casting the gel

There are many commercially available sets of apparatus for casting (or pouring) and running agarose gels (*Figure 2.5*). Most suppliers

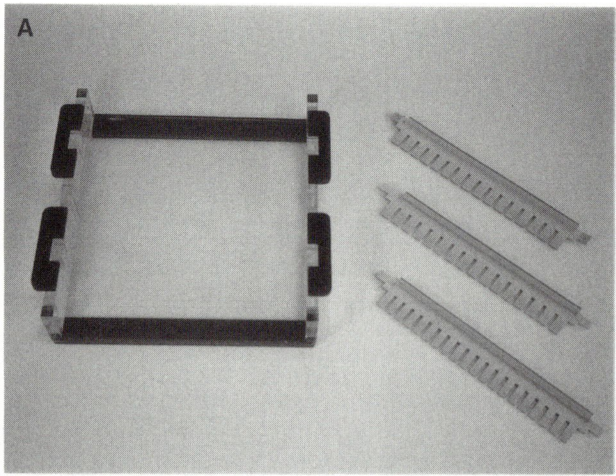

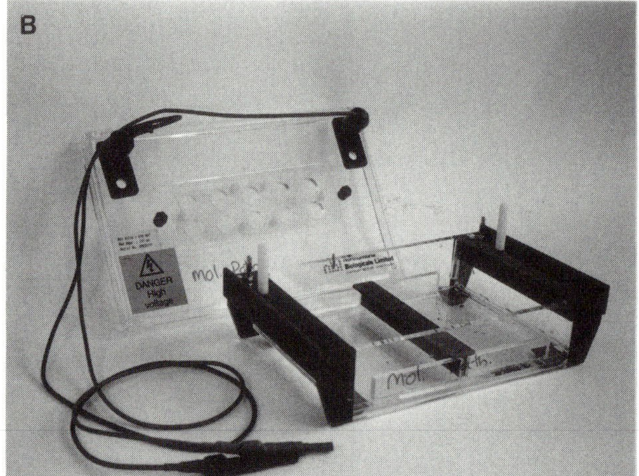

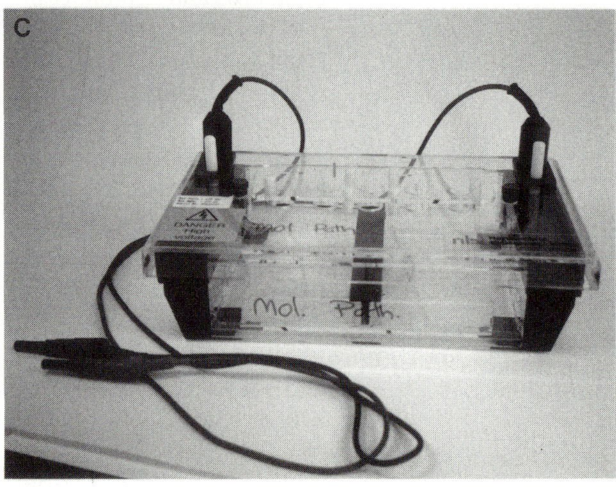

Fig 2.5

A. A plastic gel-casting tray and three combs for forming wells. **B**. An agarose gel electrophoresis tank with its lid off. **C**. An agarose gel electrophoresis tank with its lid on. For safety, electricity is supplied to the tank through terminals in the lid. Electrophoresis can only be performed when the lid is on

have by now developed systems that are safe, robust, and easy to use. Nevertheless, casting the gel provides a wonderful opportunity for all sorts of problems to occur. These are just a few that we have experienced:

- the gel sets before it has been poured,
- molten gel leaks from the casting tray while setting,
- air bubbles form around the wells,
- the set gel is too thick or too thin,
- the comb has touched the bottom of the casting tray during setting so that the wells have holes in the bottom,
- the gel is made up in the wrong buffer.

These are all quite easily avoided.

First, prepare the plastic casting tray (sometimes called the gel former or gel mould). These have a plastic base and two sides but are generally open-ended. The ends must therefore be taped up (*Figure 2.6*). Before doing this, make sure the tray is clean and dry. The main cause of leaking is that the tape does not seal the ends properly because they were wet when it was put on. Also check that the casting tray fits the gel electrophoresis tank that you intend to use. If the casting tray has been abused in the past it may have warped to such an extent that it no longer fits the tank for which it was designed. Moreover, if equipment from different manufacturers is being used in the lab, some trays that look as if they will fit may not do so. If all is well, tape the ends of the casting tray, using tape that can withstand both liquid and the heat of the molten agarose. Autoclave tape and some brands of masking tape are suitable. Tape the ends of the tray securely so that there is no chance of leakage and make sure that the height of the tape above the base of the tray is greater than the thickness of the gel that you intend to pour. Finally, place the tray on a level surface and position the comb (*Figure 2.7*).

◇ Sellotape is not suitable. It will fall off as soon as the molten agarose is poured into the tray.

◇ A small spirit level is a useful addition to the lab.

The amount of molten agarose you should make obviously depends on the size and thickness of gel required. A typical casting tray measures 11 × 14 cm, and a typical gel is 8 mm thick. To pour such a gel, you would need to make $11 \times 14 \times 0.8 = 123.2$ ml of molten agarose. On the other hand, 100 ml of molten agarose would give a gel 6.4 mm thick in a tray of this size. The thickness of the gel should be between 5 and 8 mm. If the gel is too thin, the wells will not be deep enough to take a good-sized sample and the gel will be too fragile to handle easily. A gel that is too thick wastes agarose and slows down electrophoresis. When you have decided how much gel to prepare, weigh out the required amount of agarose and add 0.9 volumes of de-ionized water and 0.1 volume of 10 × TBE (or 0.98 volumes of water and 0.02 volumes of 50 × TAE). It is easiest to dissolve the agarose in a microwave oven, but make sure that the container is not sealed, contains no metal parts and is not over-filled.

◇ A wide-necked conical glass flask is suitable.

Molten agarose has a tendency to boil over. Since it takes only about 2 minutes on a high power setting for the agarose to dissolve completely, we suggest you keep a careful eye on the gel as it is microwaved. The critical moment is usually just after the gel appears to have stopped boiling. After a few seconds of calm, the solution can

◇ This is embarrassing, and it takes ages to clean up the mess.

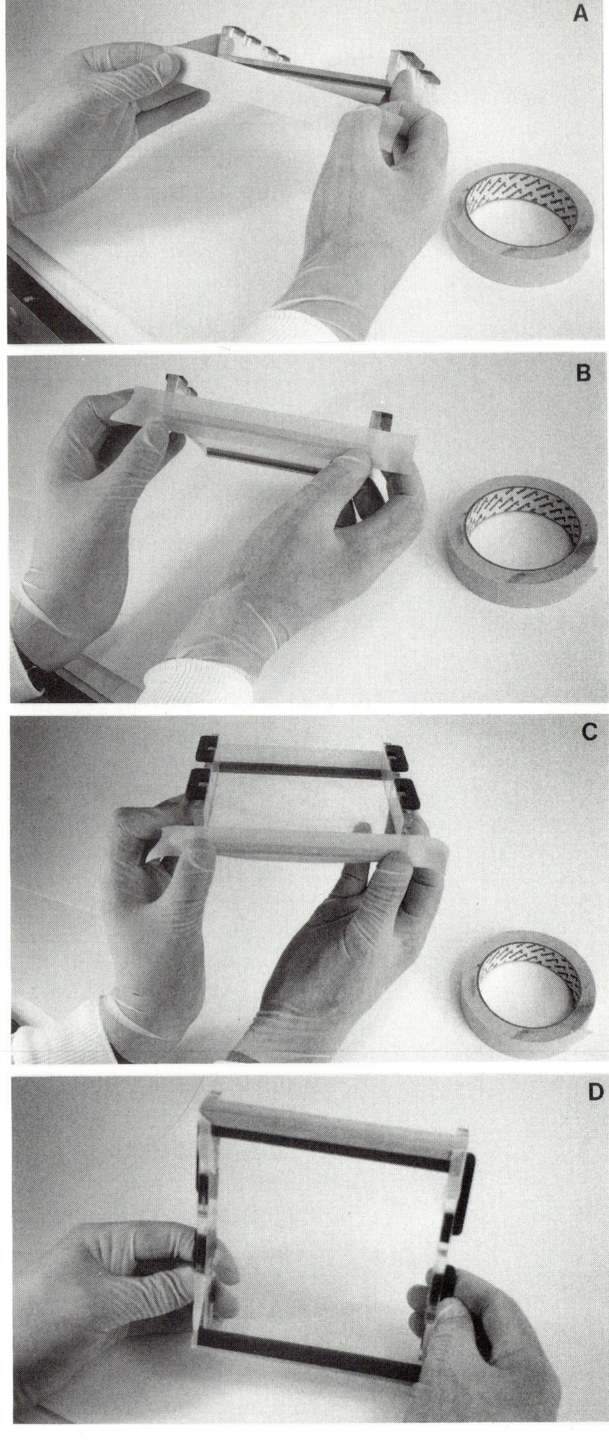

Fig 2.6

Taping the ends of the gel. **A**. Make sure that the ends of the gel tray are clean and dry before applying the tape. **B**. Apply the tape smoothly, without creases. **C**. Make sure that the edges and corners are watertight. **D**. The finished article

◇ Since there is a risk of molten agarose erupting over you, you ***must*** wear asbestos gloves, a lab coat, and eye protection and take care, as indeed you should at all times in the lab.

erupt without warning. We suggest you interrupt the power every 10–15 seconds, remove the solution, gently swirl it and then return it to the microwave oven. The molten gel is ready when no solid agarose particles can be seen. You could, of course, dissolve the

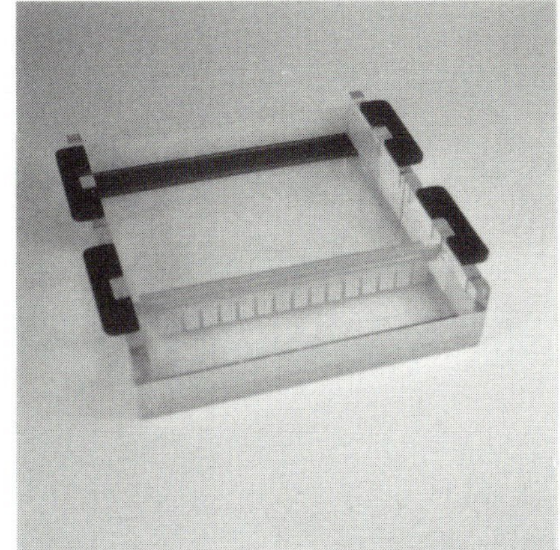

Fig 2.7

Gel tray taped at both ends, with the comb in position

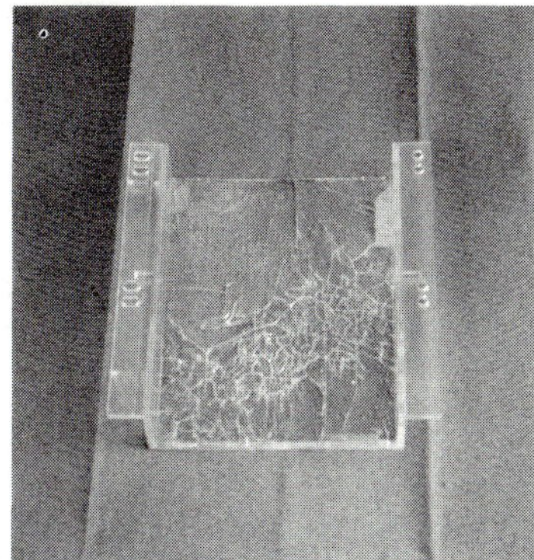

Fig 2.8

If you make a habit of pouring molten gel that is too hot, you may end up with severely cracked gel trays, like this one

agarose in a wide-necked glass conical flask over a bunsen burner. Once again, there is a risk that molten agarose will suddenly boil over. This can be avoided by gently swirling the flask during heating. This will also avoid burning the agarose powder on the bottom of the flask, which can occur if the glass becomes over-heated in the flame. Boiling the mixture can result in a significant loss of water in the form of steam. You may need to top the molten agarose solution up to the original volume with de-ionized water.

◇ If you do not notice this at the time, you will find out later when your samples flow out of the bottom of the wells and are lost.

Once the agarose has dissolved, allow it to cool to around 60 °C. If you pour the agarose solution while it is still very hot it could warp and/or crack the casting tray (*Figure 2.8*). Even minor and reversible warping can be troublesome because if the centre of the casting tray rises it can touch the bottom of the comb so that the wells form with holes in the bottom. However, take care not to let the molten agarose cool too much, or it will begin to set and you will end up with a lumpy gel that will give poor quality electrophoresis.

Immediately before pouring the gel, add ethidium bromide to a final concentration of 0.2 μg/ml, usually from a 5–10 mg/ml stock solution. Since ethidium bromide can bind to your personal DNA as well as to that in your samples, it is a carcinogen that should be handled very carefully. ***Do not*** add it to very hot agarose solutions or to the agarose mixture prior to heating since you will generate ethidium-bromide-carrying steam.

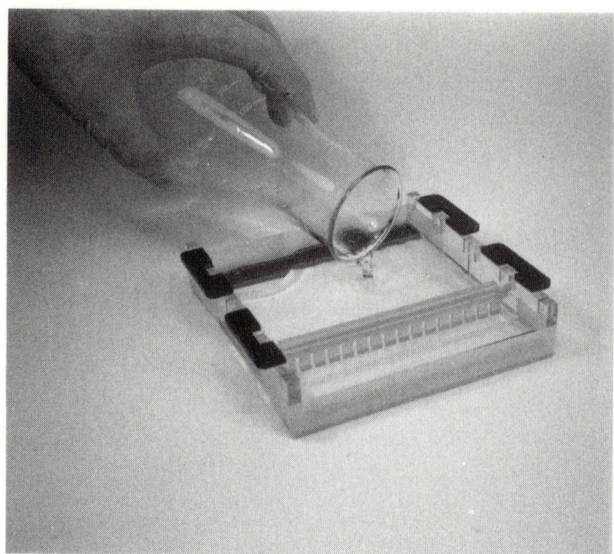

Fig 2.9

Pouring a gel

◇ You can leave the ethidium bromide out of the gel and running buffer, and stain the gel after electrophoresis by placing it in a solution of ethidium bromide (0.2 μg/ml in running buffer) for 30–60 minutes. This takes a little longer, but minimizes the volume of ethidium bromide-containing waste that you generate.

◇ The higher the concentration of agarose, the quicker the gel will set.

◇ When it has set, the gel will appear slightly opaque.

Once the ethidium bromide has been added, swirl the solution gently so that the agarose and ethidium bromide are uniformly mixed. Take care not to introduce air bubbles. Gently but confidently pour the solution into the tray (*Figure 2.9*). Check for air bubbles. These may occur anywhere but tend to congregate around the comb. Burst them, or move them out of harm's way, with a clean micropipette tip. Leave the gel to set thoroughly. If you are unsure whether it has set, **gently** tap the side of the tray. If the surface of the gel ripples, it has not set. When you are confident that the gel has set, carefully remove the tape from the ends of the tray. Take care to keep the gel level because lower percentage agarose gels, in particular, tend to slide off the tray. The gel is now ready for assembly in the electrophoresis tank.

3.4 Assembling the gel electrophoresis tank

Place the casting tray carrying the gel into the tank and carefully add running buffer (1 × TBE or 1 × TAE, depending on what is in the gel) until the surface of the buffer is 2–3 mm above the surface of the gel (*Figure 2.10*). During electrophoresis, ethidium bromide will migrate towards the negative terminal, in the opposite direction to the DNA, so that after prolonged electrophoresis there may be insufficient ethidium bromide left in the gel to allow you to see small DNA fragments. Some people therefore add 0.2 μg/ml ethidium bromide to the running buffer immediately before use. Others prefer not to spread large volumes of ethidium bromide around, since it is carcinogenic.

When you have poured the running buffer over the gel, check that air bubbles are not trapped under the casting tray. If these are not removed, the gel may not be level.

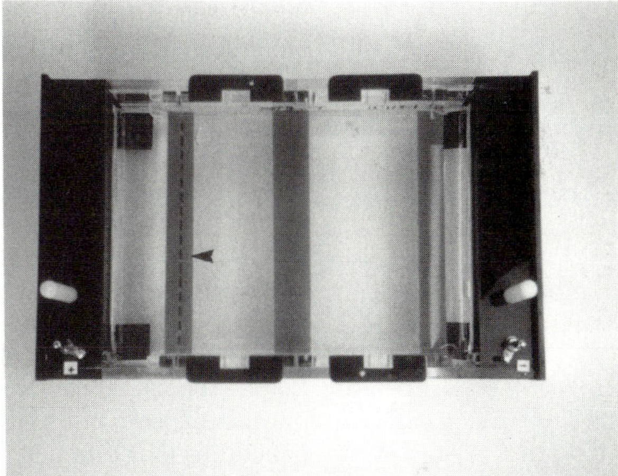

Fig 2.10

Once the gel has set, remove the tape from the ends of the tray, place the gel (still on the tray) into the electrophoresis tank, fill the tank with running buffer and remove the comb. Wells are hard to see on a white background and so this gel tray has a red strip (arrowed) on its base. If your apparatus does not have such a feature, place black paper on the bench, under the wells

The best time to remove the comb is after the addition of the running buffer. This minimizes the chances of trapping air in the wells and helps to stop the wells collapsing, which can be a problem with low percentage agarose gels. Be gentle when removing the comb as you do not want to damage the wells. If air becomes trapped in the wells, displace it by gently squirting buffer into the wells with a micropipette.

◇ It is very difficult to salvage samples once they are loaded. If your system is faulty it is best to discover this first.

At this point we suggest that you place the lid on the gel tank, connect the electrodes to the power pack and switch it on. This tests that the entire apparatus is working. If all is well, the power pack should register both a voltage and a current, and these should be of roughly the correct magnitude (see section 3.6). You should also see bubbles coming from the electrodes submerged in the running buffer and these should stop when the power pack is switched off.

3.5 Loading your samples

If you are not sure what volume of sample to put in to a well, try loading various volumes of loading buffer diluted in water. Once you

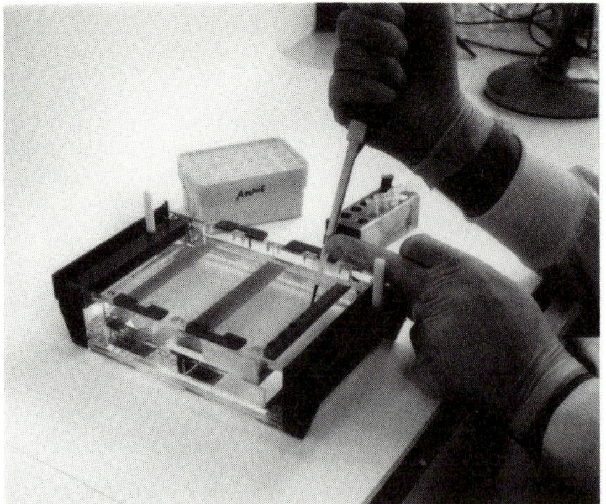

Fig 2.11

Loading a sample on to the gel. You will have better control over the pipette if you use both hands, and if you steady them on the bench, as shown here

know what volume the well will take, carefully flush the loading buffer from the well with running buffer, using a micropipette. Some people recommend that restriction-enzyme-digested DNA samples, mixed with loading buffer, are incubated at 65 °C for 2–3 minutes before loading to ensure that cohesive ends left by the restriction enzyme are properly separated. This is probably not important for enzymes that leave 4-base overhangs and is obviously unnecessary for enzymes that leave blunt ends. However, it may be worthwhile for enzymes that leave extended overhangs.

Use a micropipette to load your samples into the wells (*Figure 2.11*). Use a fresh tip for each sample to avoid cross-contamination. Loading samples is a skill that can be acquired with practice. Lower the micropipette tip *just inside* the well. If you put it in too far you risk puncturing the bottom of the well, and if you move it about too much you risk puncturing the sides. When the tip is in position, gently expel the sample into the well. Since the loading buffer is more dense than the running buffer, the sample should sink to the bottom of the well. If the sample does not sink into the well properly, it may be that the gel has been in the tank for a long time and that gel components have leached into the wells. These can be flushed out easily with running buffer.

◇ Keep your hand steady by supporting your wrist or elbow on the bench top.

◇ The dye in the sample will help you to see that it is going in the right place.

3.6 Running the gel

Once the samples have been loaded, place the lid on the tank, connect the electrodes and switch at the power pack (*Figure 2.12*). The *most important* thing is not to electrophorese the gel backwards. Remember that DNA and RNA migrate from the negative electrode towards the positive electrode.

How do you decide on what voltage to run the gel, and for how long? First, note that 'voltage' alone is a meaningless term in this context. What is important is the strength of the electric field, expressed in volts/cm, with 'cm' referring to the distance between the

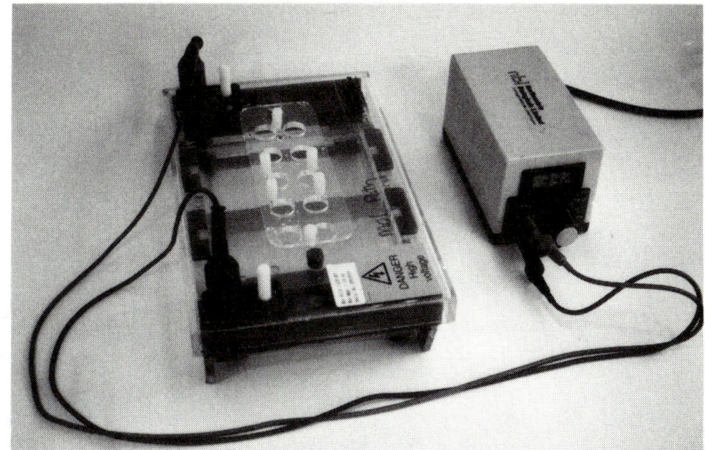

Fig 2.12

Agarose gel electrophoresis in progress

two electrodes. In weak electric fields (low volts/cm), the rate of migration of linear DNA fragments is proportional to the voltage applied. As the field becomes stronger (increasing volts/cm), the range of sizes that can be separated on a given gel decreases. Note that the greater the voltage you apply across the gel, the greater will be the current flowing through it and the greater will be the heat dissipated in the gel. If the gel becomes too hot it may begin to melt. In practice, if you want a quick result and do not mind losing resolving power in the high molecular weight range of the gel, run the gel at around 15 volts/cm. Conversely, if you need good resolution and do not mind waiting, run your gel at 3–4 volts/cm. For example, genomic DNA digests are best electrophoresed slowly at 3–4 volts/cm for 16–24 hours. This will give you good resolution of higher molecular weight fragments. In addition, because such a large amount of DNA passes from the well into the top of the gel in the initial stages of electrophoresis, fragments tend not to migrate according to size when high voltages are used. Although the fragments will sort themselves out as electrophoresis proceeds, resolution will be impaired.

◇ The heat generated is proportional to the square of the current.

◇ Gel tanks often have a maximum safe voltage written on the side of them. This *must not* be exceeded.

Use the progress of the marker dye(s) to gauge when to stop electrophoresis. For example, when electrophoresing digested genomic DNA, it is wise to run the gel until the bromophenol blue dye has migrated three-quarters of the way along the gel. This ensures that no small fragments run off the end of the gel. If you have already determined that there are no small fragments in your sample, electrophorese the gel for longer in order to improve the resolution and sizing of the larger fragments near the wells. When electrophoresing digests of cloned DNA, with fragments of known size, the gel can be viewed during electrophoresis and a decision made as to whether to electrophorese the gel any longer.

3.7 Visualizing the DNA with UV radiation

◇ Remember to wear gloves when handling gels containing ethidium bromide.

To see DNA fragments that have bound ethidium bromide, take the gel, still in the casting tray, to a transilluminator (*Figure 2.13*). You

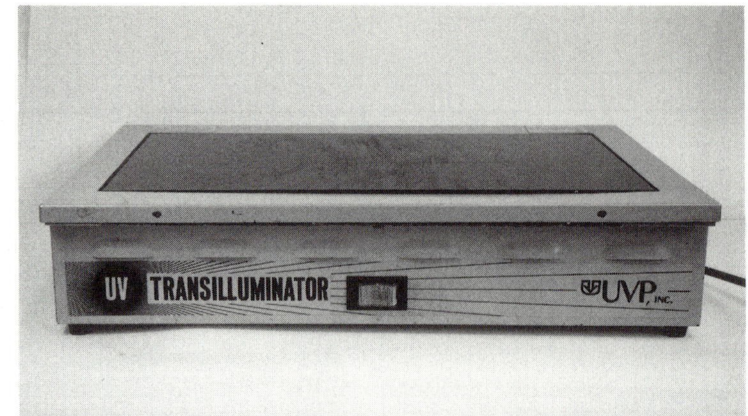

Fig 2.13

A UV transilluminator. The gel is placed on the top

◇ Wet agarose gels are slippery. To avoid fragmenting the gel on the floor (jigsaw gels do not blot well), hold the casting tray at the open ends and place it in a container for transport.

◇ Do not expose your colleagues to UV radiation.

can then view the gel in ultraviolet (UV) radiation with a wavelength of 254 nm or 302 nm. The former wavelength gives more intense fluorescence with ethidium bromide than the latter, but causes more damage to the DNA. DNA damage is not really a problem when preparing gels for Southern blotting, however. In fact, it might improve the efficiency of blotting (see Chapter 3, section 1.1.1).

Be careful when using the transilluminator because exposure to UV radiation can damage your eyes and inflame your skin. Wear a *full face mask* that has been specifically manufactured for this purpose (*Figure 2.14*). Do *not* wear goggles. Be careful not to expose your skin to the radiation for too long; the areas around your wrists and forearms are especially vulnerable to burns as they can come into close contact with the UV source and will not be protected by your gloves. When you are ready, gently slide the gel out of the casting tray on to the transilluminator and turn it on. Switch off the room light to get the best view of the DNA.

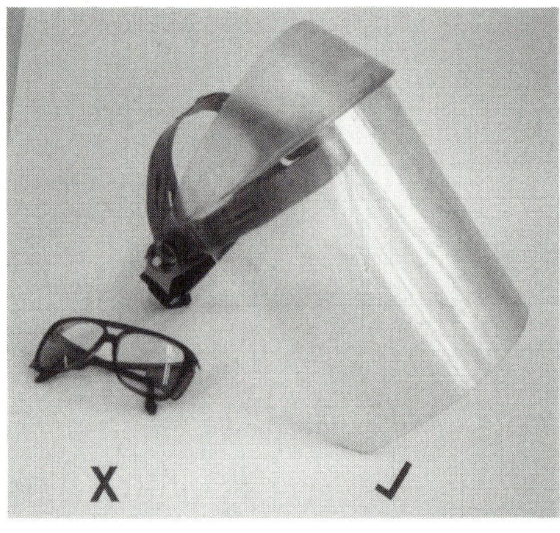

Fig 2.14

When looking at a gel on a UV transilluminator, you must use a full-face safety visor designed to block UV radiation, like the one on the right. This will protect your eyes and face. Do not use goggles, like those on the left. These will not protect your face

3.8 Photographing your gel

Gels are usually photographed by polaroid land cameras mounted above the transilluminator, although video cameras are often used nowadays.

Why should you photograph the gel? This is done not only to provide a permanent record of the appearance of the gel, although this is important, but also as an essential aid to the interpretation of the results of subsequent Southern blotting and hybridization. Photograph the gel with a ruler along one side in order to compare the positions of the DNA fragments on the photograph with those of the hybridizing fragments that you will later detect by autoradiography. Conditions for photography will vary with the illumination, the size of the gel, and the specification of the camera. Increasing the exposure time will reveal fainter bands but will also bring up the background (*Figure 2.15*). It is wise to compare the smallest bands on the photograph with the smallest bands on the gel in order to check that the photograph is an accurate representation.

◇ You can buy rulers that fluoresce in UV radiation.

◇ Align the ruler with the top of the gel or with the line of sample wells.

◇ Take time to obtain good quality photographs in which DNA and ruler are clearly visible. Photograph failures as well as successes.

◇ The photograph may not reveal all. Contaminating DNA that is invisible on a gel may be detectable when the Southern blot is hybridized.

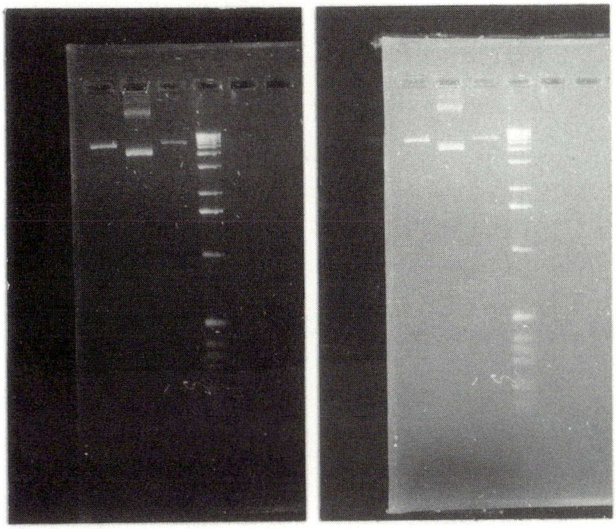

Fig 2.15

The photograph on the right has been exposed for longer than that on the left. The longer exposure reveals small DNA fragments that cannot be seen in the left-hand photograph, but the background is higher

3.9 Interpretation of gels

Examine the gel carefully for signs that your restriction digest has proceeded to completion, which means that the restriction enzyme has cut at every available recognition site.

3.9.1 Genomic DNA

Undigested and fully digested genomic DNA look very different after electrophoresis (*Figure 2.16*). Undigested DNA will form a broad band around the zone of compression towards the wells. Genomic DNA cut with an enzyme that has a 6 bp recognition site will appear as a smear of fragments from the zone of compression down to

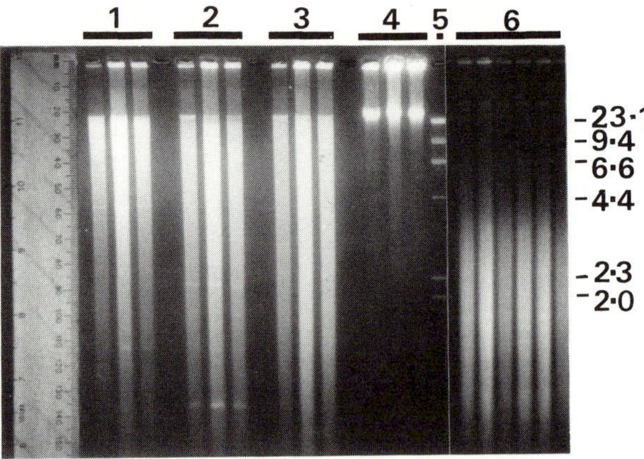

Fig 2.16

Agarose gel electrophoresis of mouse genomic DNA digested with *Bam*HI (1), *Bgl*II (2) or *Pst*I (3), which have 6 bp recognition sites, or *Rsa*I (6), which has a 4 bp recognition site. Tracks 4 contain undigested DNA. Track 5 contains λ *Hind*III size markers (sizes shown in kb). Note the bands of repetitive DNA in tracks 1 and 2

around 500 bp. Smaller fragments will also be present but will not be visible. Do not be tempted to cut this lower part of the gel off just because you cannot see any DNA there; you may be discarding fragments of interest. Genomic DNA cut with an enzyme that has a 4 bp recognition site will appear as a smear of fragments with a lower average length than that seen with a 6 bp cutter. Another useful indication of the quality of a genomic DNA digest is the presence of faint bands emerging from the background smear of DNA. These represent repetitive elements. These are sequences that occur many thousands of times in the genome and give rise to many more fragments of a particular size than would be expected if the genome contained only unique sequences. Not all restriction enzymes cut within repetitive elements in such a way as to yield such bands, however, so do not worry if you cannot see them. With experience it is possible to judge whether your genomic digest has proceeded to completion, although the final proof will lie in the autoradiograph of your hybridized Southern blot.

3.9.2 Cloned DNA

◇ The same applies to circular cosmid DNA.

Undigested and digested plasmids are also easy to distinguish. Undigested plasmids usually appear as two discrete bands on the gel (*Figure 2.17*). The faster-migrating band is supercoiled circular plasmid DNA, which is the natural state of intact plasmid DNA. The slower-migrating band is relaxed (or open) circular plasmid DNA. This arises as a result of the introduction of one or more single-strand breaks, or nicks, into the plasmid DNA during its extraction from bacteria. If a circular plasmid DNA molecule is cut by a restriction enzyme at one site it becomes linear and usually migrates between the relaxed and supercoiled circular forms (Thorne 1967). Only the linear form migrates at a position corresponding to its length in base pairs. Supercoiled DNA migrates faster than would be expected from its length because the supercoiling makes the molecule more compact and so reduces the physical resistance that it encounters as it passes through the gel. Conversely, relaxed circular DNA molecules are bulkier than linear molecules containing the same number of base

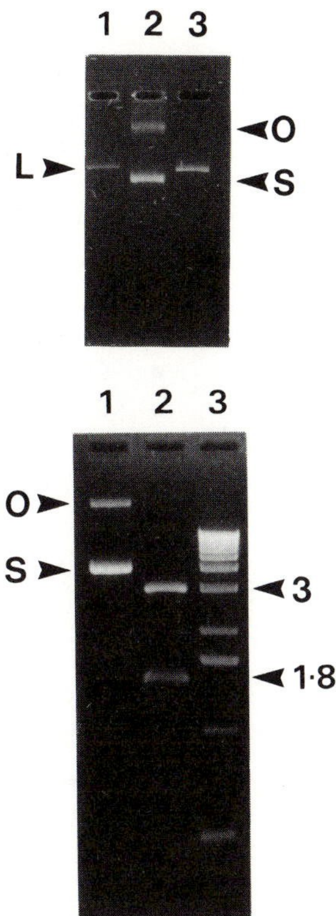

Fig 2.17

Agarose gel electrophoresis of uncut plasmid DNA (track 2) and of the same plasmid, linearized by digestion with *Bam*HI, which cuts it once (tracks 1 and 3). Supercoiled (S), open circle (0), and linear (L) forms of the plasmid are indicated

Fig 2.18

Agarose gel electrophoresis of plasmid DNA. Supercoiled (S) and open circle (O) forms of uncut plasmid (track 1) are indicated. Digestion with *Pvu*II, which cuts the plasmid at two sites (track 2) yields fragments approximately 3 kb and 1.8 kb. Track 3 contains '1 kb DNA ladder' size markers (Gibco-BRL, catalogue number 520-5615SA)

pairs and so encounter greater physical resistance as they pass through the gel, making them migrate more slowly.

Since undigested plasmid DNA samples contain multiple bands, they can be mistaken for digested samples in which an insert fragment has been successfully cut out (*Figure 2.18*). This confusion can be avoided if you adopt the habit of running undigested samples on the same gel as digested samples. This will also enable you to spot any uncut supercoiled or relaxed circles in your digested sample, indicating that digestion has been incomplete.

◇ The same applies to linear bacteriophage λ DNA and its digestion products.

If a restriction enzyme cuts a plasmid into several fragments and if digestion has gone to completion, the fragments will be present in equimolar amounts. Since DNA fragments bind increasing amounts of ethidium bromide in proportion to their length, larger fragments should fluoresce more brightly under UV radiation than smaller fragments. If you can see bands that shine less brightly than smaller bands, these represent partially digested fragments present at low concentration. The presence of such fragments indicates that digestion is incomplete.

3.10 What can go wrong with electrophoresis?

A review of the mistakes than can be made up to this point makes sober reading. Each of these has been made at least once by at least one of the authors and series editors.

- Forgetting to put in the agarose.
- Forgetting to put in TAE or TBE buffer.
- Exploding the agarose in the microwave oven.
- Warping the casting tray by pouring the gel while it is too hot.
- Failing to include ethidium bromide in the gel.
- Failing to seal the ends of the tray properly with tape.
- Removing the tape too early.
- Putting 10 × TBE, instead of 1 × TBE, into the gel tank.
- Failing to test the equipment before loading the gel.
- Overloading the gel.
- Failing to include undigested DNA or size markers.
- Running the gel backwards.
- Running the gel for too long and losing fragments of interest.
- Dropping the gel while transporting it.
- Failing to photograph with a ruler.

So do not worry if the gel does not work perfectly first time!

4. Further reading

Ogden, R.C. and Adams, D.A. (1987). Electrophoresis in agarose and acrylamide gels. *Methods in Enzymology*, **152**, 61–87.

Perbal, B. (1988). *A practical guide to molecular cloning*, (2nd edn), pp. 340–9, 356–60. John, New York.

◇ Describes pulsed-field gel electrophoresis (pp. 6.50–6.59).

Sambrook, J., Fritsch, E.F., and Maniatis, T. (1989). *Molecular cloning: a laboratory manual*, (2nd edn), Vol. 1, pp. 6.3–6.35, 9.32–9.33 6.50–6.59. Cold Spring Harbor Laboratory Press.

5. References

Bhagwat, A.S. (1992). Restriction enzymes: properties and use. *Methods in Enzymology*, **216**, 199–224.

Birnboim, H.C. (1992). Extraction of high molecular weight RNA and DNA from cultured mammalian cells. *Methods in Enzymology*, **216**, 154–60.

Burke, D.T., Carle, G.F. and Olson, M.V. (1987). Cloning of large segments of exogenous DNA into yeast by means of artificial chromosome vectors. *Science*, **236**, 806–12.

Gross-Bellard, M., Oudet, P. and Chambon, P. (1973). Isolation of high-molecular-weight DNA from mammalian cells. *European Journal of Biochemistry*, **36**, 32–8.

Guidet, F. and Langridge, P. (1992). Megabase DNA preparation from plant tissue. *Methods in Enzymology*, **216**, 3–12.

Ish-Horowitz, D. and Burke, J.F. (1981). Rapid and efficient cosmid cloning. *Nucleic Acids Research*, **9**, 2989–98.

Smith, C.L. and Cantor, C.R. (1986). Pulsed-field gel electrophoresis of large DNA molecules. *Nature*, **319**, 701–2.

Smith, M.R., Devine, C.S., Cohn, S.M. and Lieberman, M.W. (1984). Quantitative electrophoretic transfer of DNA from polyacrylamide or agarose gels to nitrocellulose. *Analytical Biochemistry*, **137**, 120–4.

Southern, E.M. (1975). Detection of specific sequences among DNA fragments separated by gel electrophoresis. *Journal of Molecular Biology*, **98**, 503–17.

Thorne, H.V. (1967). Electrophoretic characterization and fractionation of polyoma virus DNA. *Journal of Molecular Biology*, **24**, 203.

3 Southern blotting II: performing the blot

There are a number of different ways to perform a Southern blot. These include:

- capillary blotting
- electrophoretic transfer (electroblotting)
- vacuum blotting
- positive-pressure blotting.

All of these methods are suitable for blotting agarose gels. However, capillary blotting is by far the most popular method for blotting agarose gels, largely because it requires no special equipment and is quite easy to do. With some of the other methods, DNA transfer is more efficient and/or more rapid, but these advantages are marginal unless, perhaps, your laboratory has a very large throughput of blots. For these reasons, we will focus on capillary blotting, returning briefly to the other methods in section 4.

◇ Electroblotting is the only efficient method for transferring DNA from a polyacrylamide gel.

There are a number of different capillary blotting protocols. These include:

- uni-directional capillary blotting on to a single membrane,
- uni-directional capillary blotting on to multiple membranes,
- bi-directional capillary blotting.

Each of these procedures can be performed using a nitrocellulose filter or one of a wide range of nylon membranes. Blotting on to nitrocellulose filters must be performed at neutral pH, but blotting on to nylon membranes can be performed at either neutral or alkaline pH.

◇ As discussed in Chapter 7, we strongly recommend that you use nylon membranes for Southern blotting.

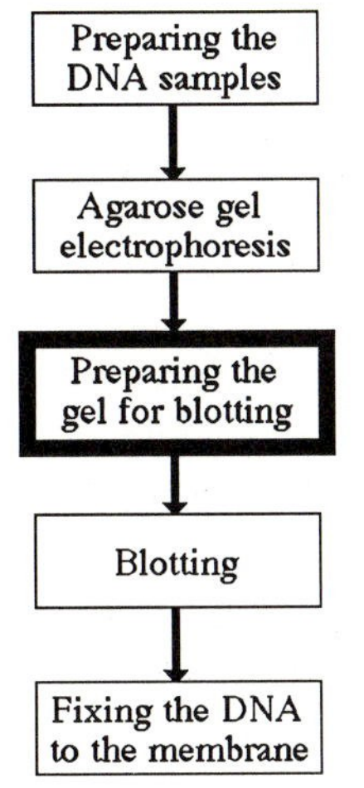

1. Uni-directional capillary blotting on to a single membrane at neutral pH

1.1 Preparing the gel for blotting

After photographing the gel, and with it still on the transilluminator, carefully cut off any unused areas with a scalpel blade. In particular, you may wish to trim off the region above the sample wells. Cut off the bottom right-hand corner of the gel to help you orientate the gel during subsequent steps. Carefully slide the gel back on to the casting tray and then slide it off the tray and into a plastic sandwich box.

Fig 3.1

An agarose gel in a plastic sandwich box containing denaturation solution

◇ Take care not to scratch the surface of the transilluminator with the blade.

◇ Take care not to drop or tear the gel.

◇ Take care not to drop or break the gel while decanting solutions.

Before you blot the gel, you must soak it sequentially in three solutions. These are:

- depurination solution,
- denaturation solution,
- neutralization solution.

Recipes for these solutions are given in the instruction manuals provided with all brands of nylon membrane. There are minor differences in the composition of the solutions recommended by different manufacturers. You should take account of these, since each manufacturer has presumably determined which solutions work best with their membranes.

Soak the gel in several volumes of each solution, in the plastic sandwich box (*Figure 3.1*). It may help to agitate the box gently, perhaps on an orbital shaker, but this is not necessary. Between each treatment, rinse the gel briefly with de-ionized water.

1.1.1 Partial depurination

In the capillary blotting method, DNA fragments are transferred out of the gel in a flow of liquid drawn through the gel by capillary action. The rate at which DNA fragments leave the gel depends on their size; the larger the fragment, the slower will be its progress. As blotting proceeds, the gel becomes dehydrated and the effective agarose concentration increases to a point at which DNA molecules can no longer leave the gel. Small fragments will have transferred to the membrane long before this point is reached, but large fragments may become trapped in the gel. To increase the rate at which large fragments leave the gel, and therefore to increase the efficiency of their transfer, the DNA can be cleaved into smaller pieces before blotting. This is done by partial depurination of the DNA in the gel.

◇ Adenine and guanine are purines.

◇ Depurination solution is typically 0.25 M HCl.

◇ Thymine and cytosine are pyrimidines.

As the term suggests, partial depurination involves removal of some of the purines from the electrophoresed DNA. This is achieved by soaking the gel for 10–15 minutes at room temperature in depurination solution. The covalent bond that connects a purine base to a deoxyribose unit in DNA is more sensitive to HCl than the bond that connects a pyrimidine base to a deoxyribose unit, and so HCl preferentially removes purines from DNA. When the gel is subsequently treated with alkali (see section 1.1.2), the phosphodiester bonds holding the backbone of the DNA strand together are cleaved at sites of depurination, resulting in fragmentation of the DNA.

There are hazards associated with depurination. If you depurinate for too long or in too strong an acid, the resulting DNA fragments will be very small and may not bind efficiently to the membrane or form stable hybrids with your probe. Even if you follow the instructions for depurination correctly, you may still encounter problems. For example, depurinated DNA can give rise to fuzzy bands, presumably because the fragmented DNA molecules are able to diffuse laterally within the gel during subsequent treatment and blotting.

◇ We have successfully blotted cloned DNA fragments of up to 40 kb without depurination, although with reduced efficiency compared to smaller fragments.

You should therefore avoid depurination whenever possible. It is only necessary when you want efficient transfer of DNA fragments greater than about 10 kb in length and is really only required when these fragments are present at low abundance, for example in a genomic digest. When you are working with cloned DNA fragments of this size, you will have relatively large amounts of each fragment and so will be able to live with a reduced efficiency of transfer and should avoid the perils of depurination.

1.1.2 Denaturation

◇ Denaturation solution is typically 1.5 M NaCl, 0.4 M NaOH.

◇ The alkali also cleaves DNA at sites of depurination.

DNA in the gel is denatured by soaking the gel for 30 minutes in denaturation solution. To ensure thorough denaturation, soak the gel sequentially in two batches of denaturation solution. The resulting single-stranded DNA will be transferred from the gel more efficiently than double-stranded DNA and, of course, is required for subsequent hybridization.

1.1.3 Neutralization

◇ Neutralization solution is typically 1.5 M NaCl, 1 M Tris-HCl, pH 7.4.

Nitrocellulose filters become extremely brittle in alkali and so it is very important to neutralize the gel after denaturation when using them. Nylon membranes are not damaged by alkali and so it is not as important to neutralize the gel before blotting when using these. Indeed, as discussed in section 3, some protocols recommend that you blot in the presence of alkali. However, blotting in alkali can lead to increased background signals after hybridization of the membranes. We therefore recommend that you neutralize the gel before blotting by soaking it in neutralization solution. To ensure thorough neutralization, soak the gel sequentially in two batches of neutralization solution. Whatever you decide to do, you *must* make sure that the membrane is neutralized before hybridization, since nucleic acid hybrids are unstable at high pH.

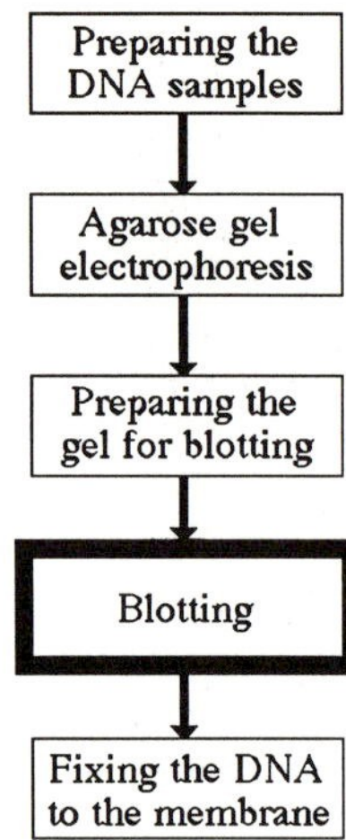

1.2 Assembling the blot

While the gel is soaking in neutralization solution, you should prepare the membrane and begin to assemble the blotting apparatus.

1.2.1 Cutting and preparing the membrane

Cut the membrane to the exact size of the gel to be blotted, using a sharp scalpel blade. Draw guide lines on the membrane with pencil, if you wish (Figure 3.3). Nitrocellulose filters and some nylon membranes must be wetted before use. Do this as described in Chapter 7.1. Most nylon membranes can be used dry.

With some nylon membranes, one surface binds DNA more efficiently than the other. Follow the manufacturer's instructions carefully to ensure that the correct side is placed next to the gel.

1.2.2 Assembly and blotting

Blots can be assembled using apparatus made for the purpose or from all sorts of bits and pieces found in the lab. We will first list the components you will need and then show you how to assemble them. Some of these components are shown in *Figure 3.2*. You will need:

◇ Molecular biologists are strangely attached to 20 × SSC (20 times concentrate of Sodium Salt Citrate). This contains 3 M sodium chloride and 30 mM sodium citrate. Make it in large quantities and store at room temperature. If a protocol calls for 10 × SSC or 1 × SSC, dilute the 20 × SSC stock 1/2 or 1/20, respectively.

◇ 20 × SSPE (Sodium Salt Phosphate EDTA) is 3.6 M sodium chloride, 200 mM sodium phosphate, pH 6.8, 20 mM EDTA.

- Transfer (blotting) solution. A high salt concentration is **essential** for efficient binding of DNA to nitrocellulose and you should use 20 × SSC to blot on to nitrocellulose filters (Southern 1975; Nagamine *et al.* 1980). High salt concentrations are **not** required for efficient binding of DNA to nylon membranes. Indeed, DNA will bind efficiently to nylon membranes in de-ionized water. Nevertheless, it is usual to use 10x SSC to blot on to nylon membranes, since salt facilitates the transfer of DNA from the gel (Khandjian 1987). Some people use SSPE in place of SSC, because it has greater buffering capacity. In practice, we have not found SSPE to have any particular advantage.
- A reservoir to hold the transfer solution. We use purpose-built plastic boxes, plastic sandwich boxes, glass developing dishes or

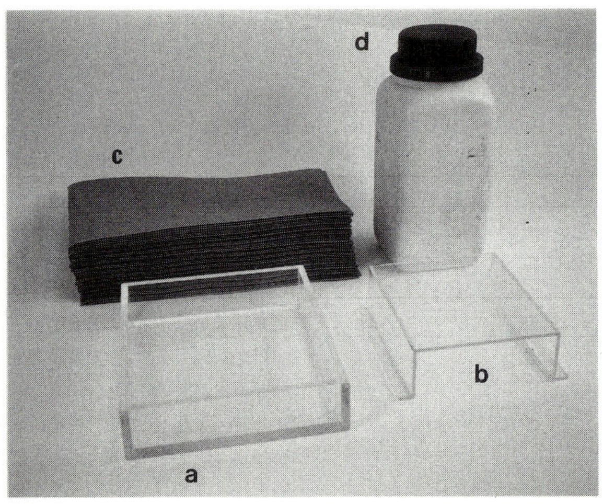

Fig 3.2

Some of the components required for Southern blotting: purpose-built perspex reservoir (**a**), purpose-built perspex support for the gel (**b**), paper towels to provide capillarity (**c**) plastic box containing 500 g of sodium chloride, for use as a weight (**d**)

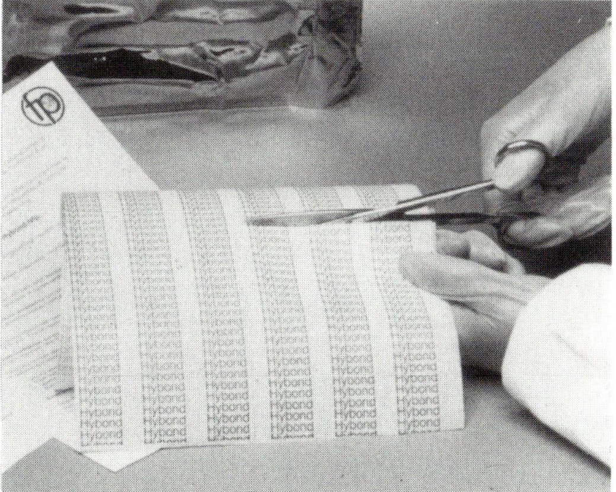

Fig 3.3

Cut nylon membrane to the correct size using scissors. Alternatively, lay it flat and cut it with a scalpel blade. Wear gloves and keep the membrane between the protective paper sheets (labelled 'Hybond'). (Photograph courtesy of Amersham International plc)

anything else that comes to hand. However, do *not* use a gel tank for this purpose, even though it is a convenient size. If you do, the salt in the transfer solution will corrode the electrodes.

- A support for the gel. We use either a purpose-built plastic platform that sits inside the reservoir or a glass plate that balances across the top of the reservoir.
- A wick of Whatman 3MM filter paper, or equivalent, to connect the reservoir and the gel. Cut this to size with scissors and pre-wet it in transfer solution. Sponges are becoming popular instead of wicks in some laboratories. Any square or oblong sponge will do, including brightly coloured ones, but they should be soaked in lots of water overnight before their first use.
- The gel.
- Saran Wrap or cling film.
- The membrane (*Figure 3.3*).
- Two sheets of Whatman 3MM filter paper, or equivalent, cut to the same size as the gel and pre-wetted in transfer solution.
- A pile of dry porous paper, approximately 10 cm high, to provide capillarity. We use paper towels.
- A weight (approximately 500 g) to keep the layers of the blotting apparatus in close contact. We use glass bottles containing water, plastic boxes containing salt and many other things.

◇ In the former Soviet Union, they used *Pravda* in blots, as well as in the toilets.

◇ Provide plenty of transfer solution so that it does not run out during blotting.

◇ Do not forget to remove the Saran Wrap before placing the gel on top of the wick.

These components should be assembled as shown in *Figures 3.4* and *3.5*. Fill the reservoir with transfer solution and put the support in place. Place the wick so that each end dips into the transfer solution and the middle lies flat on the support. Remove air bubbles between the wick and support by smoothing the wick's surface with gloved fingers or by rolling a clean pipette over the surface of the membrane. If you are too rough, you may damage the surface of the wick. To avoid this, place a piece of Saran Wrap on top of the wick while you are smoothing it down.

Place the gel with the sample wells face down and gently smooth away any air bubbles between the wick and the gel, as before. To avoid air bubbles, have a layer of transfer solution on top of the wick and gently place the gel on top.

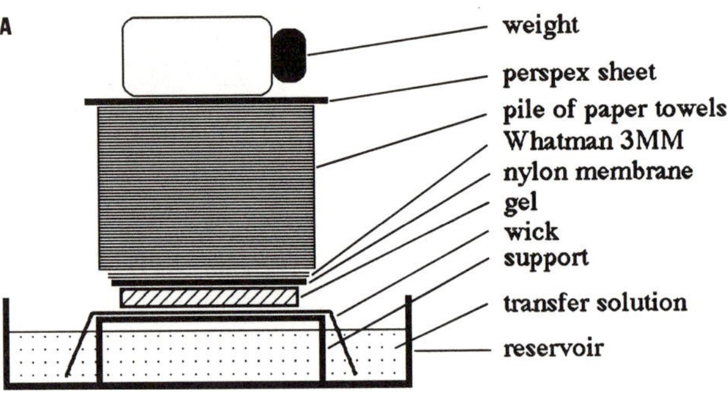

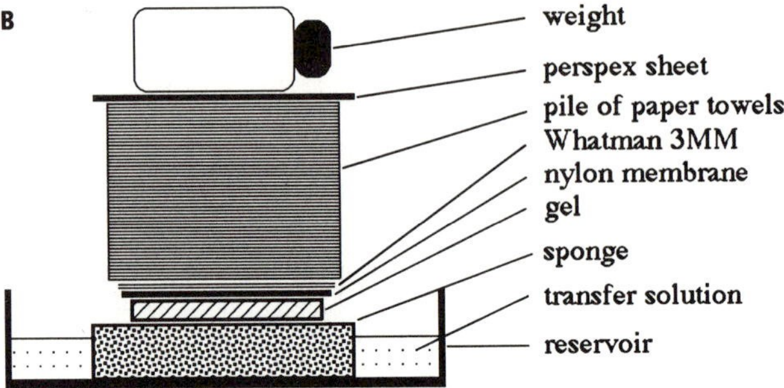

Fig 3.4

Uni-directional capillary blotting. **A.** Traditional assembly. **B.** Assembly using a sponge instead of a plastic gel support and wick

Next, place the membrane on top of the gel. To avoid air bubbles, make sure that there is plenty of transfer solution on top of the gel. Hold the edges of the membrane with gloved fingers so that the middle of the membrane bends down slightly under its own weight and gently lower it on to the gel. Make sure that the top of the membrane is correctly aligned with the top of the gel. If air bubbles do form, gently remove them as before. Using scissors, carefully snip off the corner of the membrane that corresponds to the cut corner of the gel. You must not move the membrane once it has been placed on the surface of the gel; some DNA will be transferred to the membrane almost immediately and if you move it you may end up with a 'double-exposure' effect after hybridization.

◇ If you move the membrane by mistake, discard it and use another.

Next, surround the gel and membrane with Saran Wrap. This will stop 'short circuits', in which transfer solution flows directly from the reservoir to the dry paper towels above the membrane, rather than passing through the gel. The Saran Wrap should just cover the edges of the membrane. Make sure that it does not cover too much of the membrane and so prevent the flow of transfer solution.

◇ Short circuits will dramatically reduce the efficiency of blotting.

Next, place the two pre-wetted sheets of 3MM filter paper on top of the membrane and remove any air bubbles as before. Finally, place the pile of paper towels on top of the 3MM paper, put a glass plate on top of the pile and weigh down the layers with the weight.

For convenience, the assembled blot is usually left undisturbed overnight. However, much shorter blotting times can be used. DNA fragments of less than 1 kb are transferred quantitatively within an

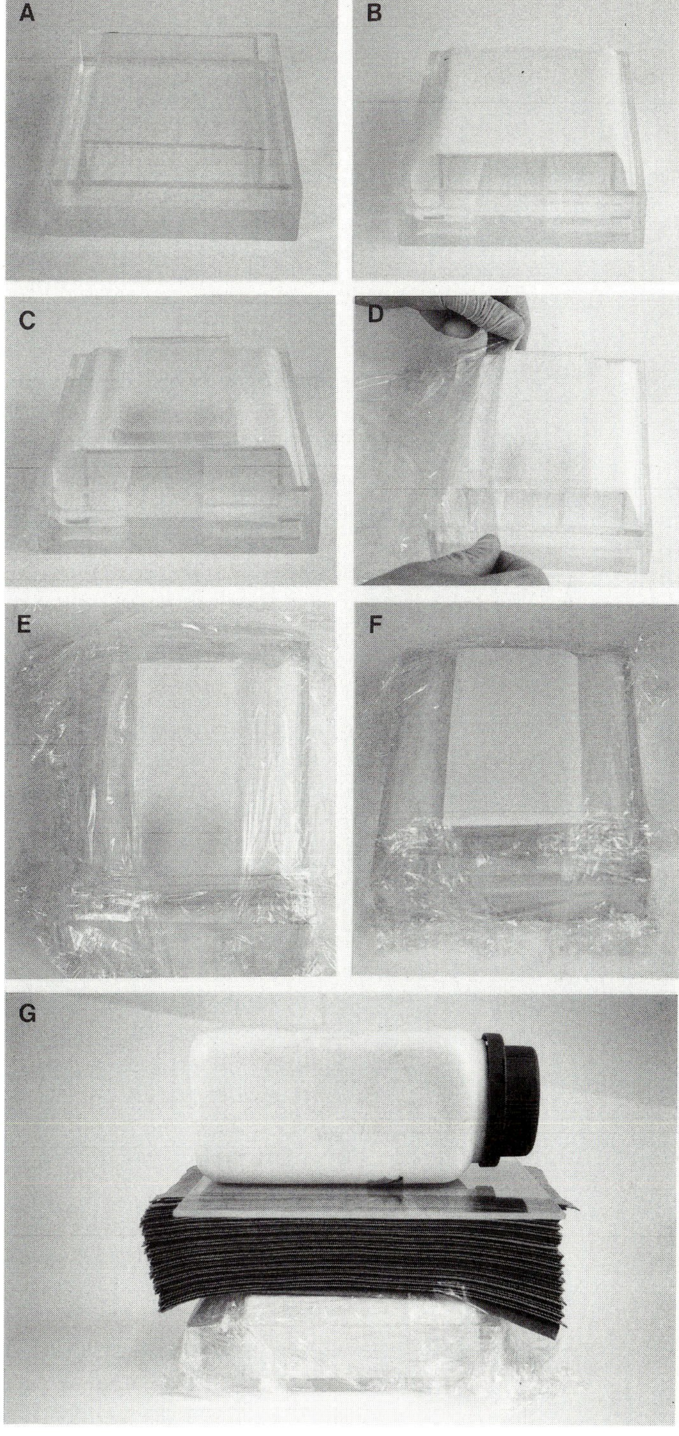

Fig 3.5

Assembling a Southern blot. **A**. Plastic box containing transfer solution and plastic gel support. **B**. Wick of Whatman 3MM filter paper in place. **C**. Agarose gel in place. **D**. Placing Saran Wrap around the gel. **E**. Saran Wrap in place. **F**. Nylon membrane in place. **G**. The complete assembly.

hour or two. If the DNA has been depurinated, even very large fragments will have been cut into smaller fragments of about this size, and so transfer is probably complete within two hours.

1.3 Dismantling the blot

◇ If you are happy that you know exactly how the filter is orientated with respect to the gel, you can carefully peel off the membrane without marking the wells.

◇ Wear gloves when handling membranes.

Gently remove the pile of paper towels (which should now be damp), the two sheets of 3MM filter paper and the cling film. Take hold of the ends of the wick and lift it up, along with the gel and the membrane. Carefully turn this over and place it, membrane side down, on to a sheet of dry 3MM filter paper. Remove the wick, without disturbing the gel, and mark the position of the wells on the membrane using a sharp soft-lead pencil or a ballpoint pen. To do this, you will need to pierce the bottoms of the wells with your pencil. Mark the positions of all the wells, whether they were used or not. You will find this extremely useful when you need to orientate the final autoradiograph and the photograph of the gel.

Gently peel off the gel and discard it. Take the membrane and rinse it briefly in a bath of 2 × SSC. Remove any pieces of agarose sticking to the membrane by gentle wiping with gloved fingers, since they will increase background hybridization. Finally, remove the membrane from the 2 × SSC, allow excess fluid to drain away and place the membrane, DNA-side up, on a piece of dry 3MM filter paper. Some manufacturers suggest you soak the membrane in denaturation solution (see section 1.1.2) for 30–60 seconds to ensure that the DNA is completely denaturated. If you do this, and we never do, you must then soak the membrane in neutralization solution (see section 1.1.3), followed by 2 × SSC.

1.4 Checking the efficiency of transfer

The only way to be certain that blotting has been successful is to include a positive control DNA sample on the gel and to detect it on the membrane later, by hybridization. However, there are a few clues that you can look for while dismantling the blotting assembly. If capillary blotting has worked properly:

- the pile of paper towels should be damp, or show signs of having been damp and then dried in the air,
- the bromophenol blue dye should have moved from the gel to the membrane,
- the gel should be much thinner than it was before blotting.

If none of these things has happened, something has gone wrong; the flow of transfer solution through the gel may have been inefficient due to a 'short circuit' in the blotting assembly, or because you accidentally covered the gel or membrane with Saran Wrap.

You could also stain the remains of the gel with ethidium bromide, to see whether any DNA remains in it. Place the gel in 0.2 μg/ml ethidium bromide in transfer solution for 30–60 minutes and view it with UV radiation. Compare what you see with the photograph of the gel before blotting. You should find, particularly with genomic Southern blots, that low molecular weight DNA is undetectable but that some high molecular weight DNA remains. If substantial amounts of DNA remain in the gel, something has gone wrong.

1.5 Fixing the DNA to the membrane

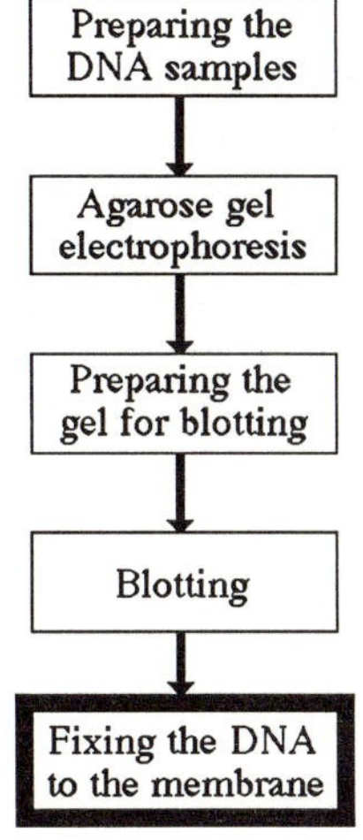

The procedure for fixing the DNA to the membrane after blotting depends on the type of membrane you are using. Follow the manufacturer's instructions closely.

In the case of nitrocellulose filters, DNA can be semi-permanently, but non-covalently, linked to the filter by baking. Sandwich the filter between two sheets of dry 3MM filter paper and bake for two hours at 80 °C. Some protocols specify the use of a vacuum oven, to minimize the risk of combustion of nitrocellulose. We have never found this necessary. You can reduce baking time to as little as 30 minutes, but should not bake for longer than two hours because nitrocellulose filters become very brittle with prolonged baking.

◇ It is not clear exactly how DNA binds to nitrocellulose.

One of the great advantages that nylon membranes have over nitrocellulose filters is that DNA can be cross-linked to them covalently, that is by the formation of covalent bonds between the DNA and chemical groups on the surface of the membrane. Such bonds are very difficult to break and so the DNA is essentially permanently linked to the membrane. There are two methods for cross-linking:

- UV treatment,
- drying.

It is very important to use the appropriate method for the particular type of membrane you are using. Follow the manufacturer's instructions closely.

1.5.1 UV treatment

◇ Nitrocellulose filters and some nylon membranes must ***not*** be exposed to UV radiation because they may catch fire.

When exposed to UV radiation, a proportion of the thymine residues in the DNA will cross-link to amine groups on the surface of the membrane (Li *et al.* 1987). The procedure is simply to wrap the damp membrane in Saran Wrap and expose the DNA-side to UV radiation. The difficulty is in deciding what dose of UV radiation to give:

- too little exposure will result in inefficient cross-linking,
- too much exposure will cross-link efficiently but will reduce the extent to which the bound DNA can hybridize, by reducing the number of thymines available for hydrogen bond formation with your probe.

The best solution to this problem is to use a UV source manufactured for the purpose of cross-linking DNA to membranes (*Figure 3.6*). This will enable you to deliver a metered dose of UV radiation in accordance with the instructions of the manufacturer of your membrane. Second best is to use a UV transilluminator (254 nm wavelength) that you have previously calibrated with a UV meter.

◇ Worst is to use a UV transilluminator and guesswork.

If you are using a UV transilluminator, it might be worthwhile testing the optimal exposure directly by blotting a gel in which each track has an identical amount of plasmid DNA, cutting the membrane into strips, cross-linking each strip with a different dose of UV

Fig 3.6

A UV cross-linker

◇ Remember to take suitable precautions when using a UV radiation source.

radiation, hybridizing the strips with a labelled probe for the plasmid, and looking for the UV exposure time that gives you the strongest hybridization signal. If you have calibrated your UV transilluminator in this way, remember that the energy that it emits will change as the bulbs age. Moreover, different bulbs in the same transilluminator are often of different ages.

1.5.2 Drying

DNA can be covalently cross-linked to some types of membrane simply by allowing them to dry. If you are not in a hurry, place the membrane in a safe place with the DNA-side up and allow it to dry thoroughly in the air. If time is short, dry the membrane thoroughly in an oven at any temperature up to 80 °C.

1.6 Storing membranes before hybridization

One the DNA has been fixed to the membrane, it can be stored indefinitely at room temperature in a dry environment. It is wise to place the membrane in an envelope of 3MM filter paper to keep dust away, and to store it in a protective box.

2. Capillary blotting on to multiple membranes at neutral pH

You are quite likely to want to probe the same set of restriction-digested DNA samples with a number of different hybridization probes. This can be done sequentially by hybridizing the membrane with one probe and then, after autoradiography, removing the bound probe ('stripping the membrane'). This allows you to re-hybridize the membrane with a second probe, and so on. However, a more rapid approach is to blot the same gel on to several membranes and to hybridize each membrane with a different probe. There are two ways in which you can blot on to more than one membrane:

- uni-directional capillary blotting on to several membranes,
- bi-directional capillary blotting.

2.1 Uni-directional capillary blotting on to several membranes

◇ Do ***not*** attempt this with genomic Southern blots

Assemble the blot exactly as described in section 1.2.2. After 5 minutes, carefully remove the pile of paper towels and the two sheets of 3MM filter paper, and peel off the membrane. Replace this with a second membrane and replace the 3MM filter paper and paper towels. After a further 10 minutes, replace the second membrane with a third. After 15 minutes, replace the third membrane with a fourth. Allow blotting to continue overnight and then remove the fourth membrane. Rinse each membrane in 2 × SSC and fix the DNA to it as soon as it is removed. This procedure works very well when you are blotting cloned DNA, because the concentration of DNA in the gel will be relatively high. The procedure is not suitable when you are blotting genomic DNA, because there is so little of each species of DNA fragment present that you want all of the DNA in the gel to be transferred to a single membrane.

2.2 Bi-directional capillary blotting

◇ Again, this works well when you are blotting cloned DNA but is inappropriate for blotting genomic DNA.

Prepare the gel as described in section 1.1, soak it for 30 minutes in 10 × SSC and assemble the blot as shown in *Figure 3.7*. Lay 10 dry paper towels on the bench, followed by a sheet of damp 3MM filter paper, a nylon membrane (cut to size and pre-wetted if called for by the manufacturer's instructions), the gel, a second nylon membrane, a sheet of damp 3MM paper, 10 dry paper towels, and a weight (500 g). As before, make sure that there are no air bubbles between the layers. Capillary action should draw the water contained in the gel in both directions, so that DNA is moved on to both membranes. After two hours, the membranes should be removed and processed as before.

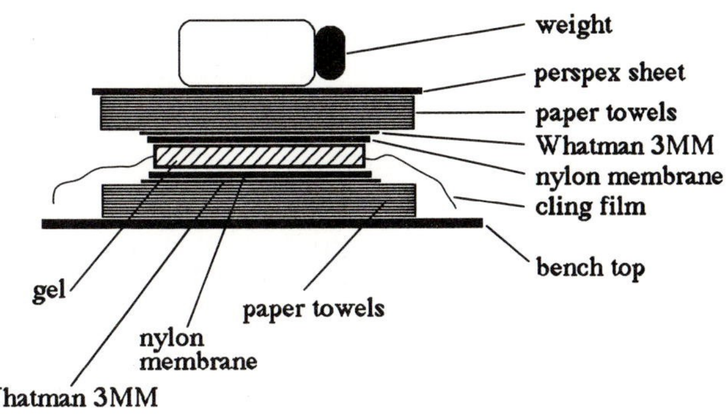

Fig 3.7

A bi-directional capillary blot

3. Capillary blotting at alkaline pH

◇ Some repeated sequences might renature during blotting.

◇ Alkaline transfer solution is typically 0.4 M NaOH, 1.5 M NaCl.

Some nylon membranes will bind DNA efficiently at alkaline pH (Reed and Mann 1985). Blotting in alkaline transfer solution can be worthwhile if the DNA you are interested in is likely to renature during blotting. Otherwise, it is best to avoid alkaline blotting since it tends to result in higher background signals after hybridization. To blot at alkaline pH, soak the gel in depurination solution and denaturation solution as before, but omit the neutralization step (see section 1.1). Blot the gel as before, using alkaline transfer solution instead of SSC. Some brands of paper towels are not resistant to alkaline transfer solution and make a nasty mess. You should check your brand before use.

After blotting, neutralize the membrane by soaking it for 15 minutes in neutralization solution before soaking in 2 × SSC and processing as before.

4. Other methods of blotting

A number of other methods for blotting DNA and RNA have been developed, including electroblotting, vacuum blotting, and positive-pressure blotting.

4.1 Electrophoretic transfer (electroblotting)

In this procedure, the gel is treated as for capillary blotting, sandwiched with a nylon membrane between porous pads, and mounted vertically in a large tank filled with buffer (*Figures 3.8* and *3.9*). An electric field is applied across the tank and DNA fragments migrate

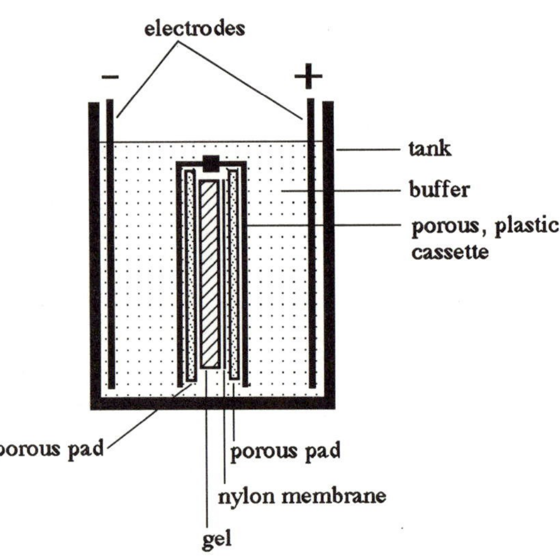

Fig 3.8

Electroblotting apparatus

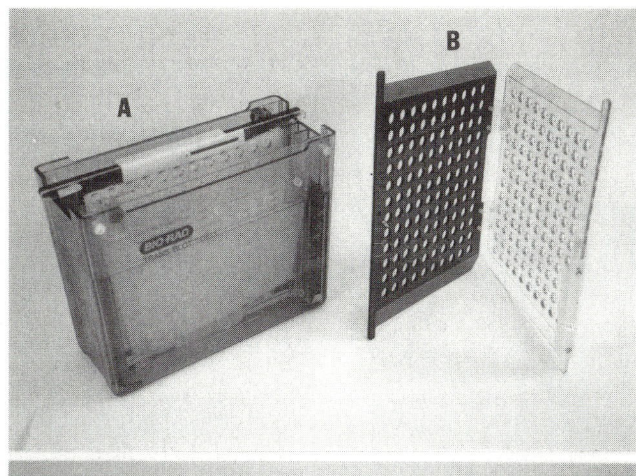

Fig 3.9

Electroblotting apparatus. A. Electrophoretic transfer cell (or tank). In the tank are the transfer buffer and the assembled cassette containing porous pads, gel, and nylon membrane. B. Cassette before assembly. C. Lid of cell, with electrical leads

towards the anode, out of the gel and on to the membrane. The advantages of electrophoretic transfer are that high molecular weight DNA fragments are transferred efficiently without the need for partial depurination and that transfer is rapid.

Electroblotting is not suitable for blotting on to nitrocellulose filters because of the high concentration of salt required for efficient binding of DNA to nitrocellulose. Buffers containing high concentrations of salt conduct electricity very well. The high electric current results in buffer exhaustion, due to electrolysis, and in heating to temperatures that interfere with the binding of DNA to the filter. To overcome these problems, large volumes of buffer must be used and the apparatus must be cooled. The inconvenience of this outweighs the advantages.

◇ The heat generated is proportional to the square of the current.

◇ There are a number of commercially available electroblotting tanks, some with cooling devices and some that must be used in a cold-room.

Electroblotting can be used to transfer DNA on to nylon membranes, since these bind DNA in solutions containing low concentrations of salt. Nevertheless, it is still necessary to cool the apparatus during blotting.

More recently, semi-dry electrophoretic transfer cells have been developed (*Figures 3.10* and *3.11*). In these, the gel and membrane are sandwiched between layers of filter paper saturated in buffer, which act as reservoirs during blotting. These are, in turn, sandwiched between two large, flat carbon or platinum electrodes. Such equipment generates relatively little heat, removing the need for cooling.

On balance, we feel that there is no great advantage in using electoblotting, rather than capillary blotting, to transfer DNA from an agarose gel. Electroblotting is the method of choice, however, for

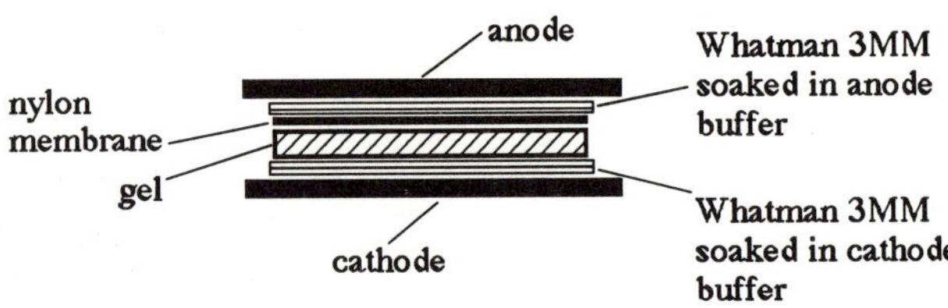

Fig 3.10

Semi-dry electroblotting apparatus

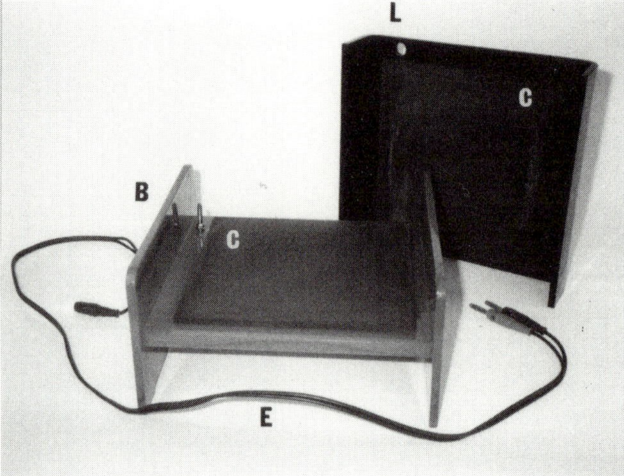

Fig 3.11

Semi-dry electrophoretic transfer cell. The base (**B**) and lid (**L**) have flat carbon electrodes (**C**), which are connected to plug-in electrical leads (**E**)

transferring DNA from a polyacrylamide gel, since capillary blotting from such gels is very inefficient.

If you are using electroblotting equipment, follow the instructions provided by the manufacturers of the equipment and of the membrane you are using.

4.2 Vacuum blotting and positive-pressure blotting

In vacuum blotting (*Figure 3.12*), the gel is treated as for capillary blotting and placed in contact with a nylon membrane or nitrocellulose filter lying on a porous support over a vacuum chamber. Transfer solution is drawn through the gel from an upper reservoir, carrying DNA fragments on to the membrane (Peferoen *et al.* 1982). There are a number of commercially available sets of vacuum blotting apparatus.

◇ If you are using this equipment, follow the instructions provided by the manufacturer.

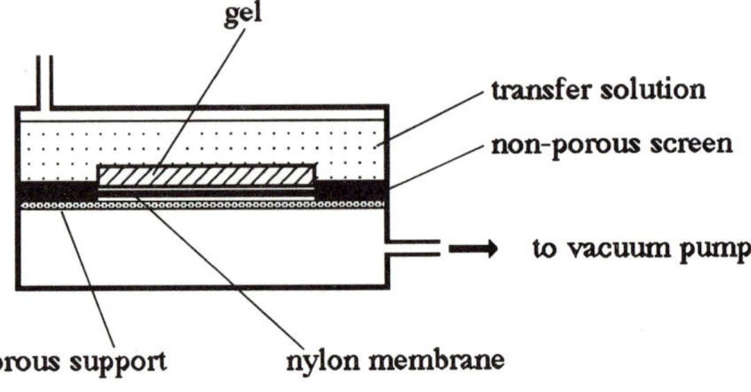

Fig 3.12

Vacuum blotting apparatus

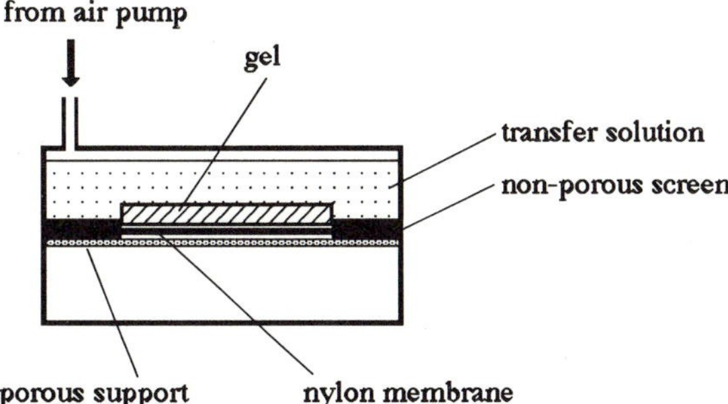

Fig 3.13

Positive-pressure blotting apparatus

Positive-pressure blotting apparatus (*Figure 3.13*) typically comprises two chambers, separated by a porous support. The gel is treated as for capillary blotting and placed on top of a nylon membrane or nitrocellulose filter lying on the porous support. The upper chamber is filled with transfer solution and sealed. Transfer solution is then forced down through the gel by pumping air into the upper chamber under pressure. There are a number of commercially available sets of positive-pressure blotting apparatus.

The main advantage of both vacuum blotting and positive-pressure blotting is that they are very rapid, with quantitative transfer of partially depurinated DNA being achieved within 30 minutes from standard agarose gels. Positive-pressure blotting avoids the problem of gel collapse, which is associated with vacuum blotting, and so is claimed to result in more rapid and efficient transfer. In practice, there is probably not much in it. The main disadvantage of these two methods is that they require equipment that is expensive, can go wrong, and needs careful use.

4.3 Which method of blotting should you use?

The answer to this question depends on the use to which you intend to put blotting and on the size of your budget. In our laboratory, a lot of people are performing a few blots and we have never felt the need to buy special blotting equipment. We have flirted with some demonstration models, but remain wedded to the capillary blot because it is easy, cheap, and trouble-free. If, however, your work requires you to perform a large number of blots, perhaps as part of a gene mapping or diagnostic exercise, it might be worth buying equipment that will enable you to turn over a large number of blots each day. We suggest that you contact manufacturers of semi-dry electrophoretic transfer cells, vacuum blotting equipment, and positive-pressure equipment, and see who is most persuasive.

5. Further reading

Perbal, B. (1988). *A practical guide to molecular cloning*, (2nd edn), pp. 423–38, Wiley, New York.

Sambrook, J., Fritsch, E.F., and Maniatis, T. (1989). *Molecular cloning: a laboratory manual*, (2nd edn), Vol. 2, pp. 9.31–9.46, Cold Spring Harbor Laboratory Press.
Wahl, G.M., Meinkoth, J.L., and Kimmel, A.R. (1987). Northern and Southern blots. *Methods in Enzymology*, **152**, 572–81.

6. References

Khandjian, E.W. (1987). Optimized hybridization of DNA blotted and fixed to nitrocellulose and nylon membranes. *BioTechnology*, **5**, 165–7.
Li, J.K., Parker, B., and Kowalin, T. (1987). Rapid alkaline blot-transfer of viral dsRNAs. *Analytical Biochemistry*, **163**, 210–18.
Nagamine, Y., Sentenac, A., and Fromageot, P. (1980). Selective blotting of restriction DNA fragments on nitrocellulose membranes at low salt concentrations. *Nucleic Acids Research*, **8**, 2453–60.
Peferoen, M., Huybrechts, R. and De Loof, A. (1982). Vacuum blotting: a new, simple and efficient transfer of proteins from sodium dodecyl sulfate-polyacrylamide gels to nitrocellulose. *FEBS Letters*, **145**, 369–72.

◇ This paper describes the vacuum blotting of proteins. The same principle works for nucleic acids.

Reed, K.C. and Mann, D.A. (1985). Rapid transfer of DNA from agarose gels to nylon membranes. *Nucleic Acids Research*, **13**, 7207–21.
Southern, E.M. (1975). Detection of specific sequences among DNA fragments separated by gel electrophoresis. *Journal of Molecular Biology*, **98**, 503–17.

Electrophoresis of RNA and northern blotting

As discussed in the previous chapter, Southern blotting is used to detect specific DNA molecules that have been size-separated by gel electrophoresis. Following the development of Southern blotting, a number of workers sought to modify the technique so that it could be used to detect specific RNA molecules that had been size-separated by gel electrophoresis (Alwine *et al.* 1977). The resulting procedure was named northern blotting, not because by some remarkable coincidence it was developed by a person named Northern, but rather as a kind of molecular biologists' joke. Probably the most important thing to know about northern blotting, therefore, is that its name begins with a lower case 'n', since it is a technique and not a person's name. If you submit a manuscript that refers to 'Northern blotting', a zealous referee or copy editor is certain to object.

◇ This tells you something about molecular biologists

◇ This tells you something about copy editors. Referees, of course, are always right.

1. How does northern blotting differ from Southern blotting?

The modifications used in northern blotting take account of differences in the physical properties of DNA and RNA. The most important of these is that RNA molecules are single stranded rather than double stranded. This results in:

- Intramolecular base pairing between short regions of complementary sequence, causing the RNA molecule to fold up (*Figure 4.1*). The rate of migration of a folded RNA molecule in an

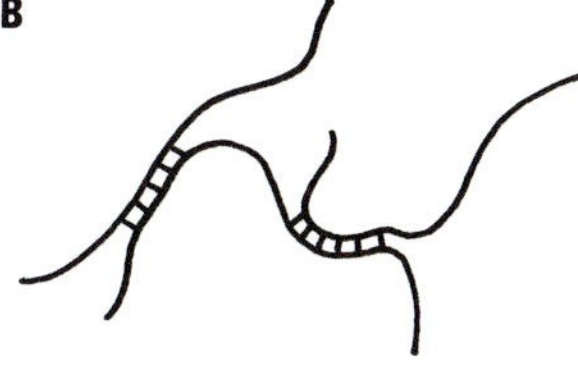

Fig 4.1

Intramolecular (A) and intermolecular (B) base pairing in RNA agarose gel will no longer depend solely on its length, but will also depend on the extent of folding; the more compact the molecule, the easier it will negotiate the pores in the gel matrix and the faster it will migrate.

- Intermolecular base pairing between complementary sequences in different RNA molecules. This will cause aggregation of the RNA (*Figure 4.1*).

RNA molecules must therefore be completely unfolded during gel electrophoresis, by the addition of suitable denaturing agent.

◇ For example, RNases can withstand boiling, whilst deoxyribonucleases (DNases), enzymes that degrade DNA, are usually inactivated by boiling.

A second important difference between RNA and DNA is that RNA is much more prone to degradation by enzymes than DNA. An added complication is that ribonucleases (RNases), the enzymes that degrade RNA, are extraordinarily stable. RNases can contaminate RNA samples both because they are present in the cells or tissues from which the RNA was isolated (endogenous RNases) and because they are introduced from outside sources, such as your fingers (exogenous RNases). Protocols for isolating RNA from cells and tissues all contain steps designed to inactivate endogenous RNases. When it comes to RNA electrophoresis and northern blotting, your task is to minimize contamination with exogenous RNases. How can you do this? It is very important to wear clean laboratory gloves at every stage of the procedure, not just during sample preparation (when the RNA is most vulnerable) but also at subsequent stages. Use gel electrophoresis and blotting apparatus that have been thoroughly cleaned in distilled water. Some laboratory manuals recommend that containers and solutions used to make and run gels are treated with diethyl pyrocarbonate (DEPC), which is an inhibitor of RNases. In our experience this is unnecessary. However, if you want to do this, proceed as follows:

◇ Plastic gloves are not magical. If they come into contact with RNases they will themselves become sources of contamination.

◇ DEPC is toxic, and may be carcinogenic. Handle it with care. DEPC reacts with ammonium ions to give ethyl carbonate, a known carcinogen. Some protocols tell you to stain RNA gels with ethidium bromide in ammonium acetate soultion. Do **not** do this if you have used DEPC to prepare the gel.

◇ You can DEPC-treat your electrophoresis tank, but do **not** autoclave it!

(1) Plasticware—fill the item with 1 per cent (v/v) DEPC in water, leave for 2 hours at 37 °C, discard the fluid and autoclave the item to remove traces of DEPC.

(2) Solutions—add 1 per cent (v/v) DEPC, leave for at least 12 hours at 37 °C and autoclave. DEPC reacts with amines, so do not use it to treat solutions of chemicals with amine groups, such as Tris-buffer.

(3) RNases on glassware can be inactivated by baking at 180 °C for at least 8 hours.

(4) Sterile, disposable plasticware is RNase-free.

A third difference between RNA and DNA is that RNA is much more sensitive to hydrolysis by acid or alkali. This makes it particularly important to use solutions with adequate buffering capacity when handling RNA. In addition, as we will see, some procedures involving treatment with acid or alkali are modified when using RNA.

2. What information can a northern blot give?

Northern blotting allows you to determine whether a particular gene probe hybridizes to one or more RNA species in a given cell type. It also allows you to estimate the length(s) of the hybridizing RNA species and gives an indication of their abundance (*Figure 4.2*). Of course, there is a limit to the sensitivity of the technique and the absence of a hybridization signal does not prove that a particular RNA is completely absent from a cell. It is also very difficult to use northern blotting data to determine the absolute amount of a particular RNA species in a cell, and the technique is rarely used for this purpose. However, the technique is widely used to make broad comparisons between the abundance of a particular RNA species in different cell types.

◇ This is fraught with difficulty, as is discussed in section 3.

An error made in a disturbing number of published papers is the idea that northern blotting provides a measure of the rate at which a particular gene is transcribed. It does not. Northern blotting simply provides a measure of the levels of an RNA species in a particular cell or tissue type at a particular time point. These levels are determined by both the rate of transcription and the rate of RNA degradation. Thus, if northern blotting reveals that an RNA species is 10 times more abundant in one cell type than in another, this could indicate any of the following:

- the corresponding gene is transcribed more efficiently in the first cell type,
- the RNA species is more stable in the first cell type,
- that both increased transcription and increased stability act to increase the levels of the RNA in the first cell type.

Other techniques must be used to distinguish between these possibilities. For example, nuclear run-on assays can be performed to measure transcription rates (Marzluff and Huang 1985), whilst RNA stability can be measured by performing northern blotting with RNA

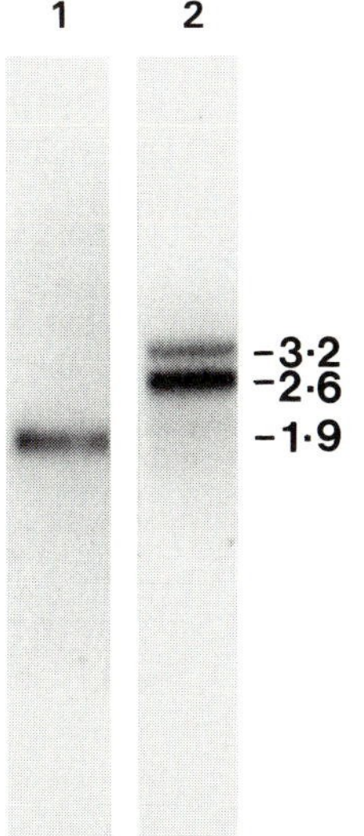

Fig 4.2

Northern blots of poly(A)-enriched mRNA (2 μg per track) isolated from chick embryo limb buds. The blots were hybridized with probes for chicken bone morphogenetic protein-4 (*Bmp-4*) mRNA (blot 1) and chicken *Bmp-2* mRNA (blot 2). Sizes of mRNAs, in nucleotides, are indicated

samples isolated from cells at intervals after treatment with actinomycin D to block new transcription. When discussing northern blotting data, it is appropriate to talk in terms of differences in RNA levels or in RNA accumulation, but not in terms of differences in transcription.

3. Comparing the levels of an mRNA species in different cell types

3.1 Equal loading

As noted above, a common application of northern blotting is to compare the levels of a particular mRNA species in different cells or tissues. However, if the hybridization signal is much more intense in one track than in another, the first thing you must do is exclude the possibility that this is simply due to unequal loading of RNA in the two tracks. This begs the question of what constitutes 'equal loading'.

◇ Messenger RNA is distinguished from other types of RNA by possession of a poly(A) tail. Polyadenylated mRNA is extracted from total RNA by oligo(dT) affinity-chromatography (Aviv and Leder 1972)

◇ Differences in the apparent abundance of RNA species introduced in this way may not even be detectable by northern blotting.

The most obvious way to get equal loading is to load the same amount of total (or polyadenylated) RNA in each track. The problem with this is that the abundance of the RNA species under study can appear to change as a result of changes in the abundance of *other* RNAs (*Figure 4.3*). However, since many (though not all) changes in RNA levels involve relatively low abundance RNA species, this effect is likely to be small in practice. Many people therefore do perform experiments of this type by loading equal amounts of total (or polyadenylated) RNA in each track.

An approach that avoids complications due to the changing abundance of other RNAs is to load the wells with amounts of RNA isolated from equal numbers of cells or equal weights of tissue. This is not without its own problems, however. For example, two pieces of tissue of the same weight may contain different numbers of live cells. An alternative is to load amounts of RNA isolated from cell or tissue samples that contain equal quantities of DNA, since DNA content provides a rough guide to cell number. If you adopt either of these alternatives, you must control for the possibility that there may be differences in the efficiency with which RNA is recovered from differ-

	Cell type A	**Cell type B**
mRNA 'X	10 copies per cell	10 copies per cell
mRNA 'Y'	10 copies per cell	1 copy per cell
All other mRNAs	80 copies per cell	80 copies per cell
Apparent abundance of mRNA X	10%	12% (approx.)

Suppose you want to know whether the abundance of mRNA 'X' is different in two types (A and B). You could load equal amounts of RNA from the two cell types on to a gel, blot it and probe the blot for mRNA 'X'. In the example shown, the abundance of mRNA 'X' appears to be greater in cell type B than in cell type A. However, it can be seen that the abundance is actually the same in both cell types, and only appears to differ as a result of a difference in the abundance of another mRNA (Y), of which you know nothing. It would be better to load amounts of RNA isolated from equal numbers of the two cell types.

Fig 4.3

Take care when assessing changes in mRNA abundance

◇ Whichever approach you use, you must be aware of its limitations and interpret your results accordingly. Most important, when reporting your work, make it clear what you have done so that others may judge the validity of your conclusions.

ent cells or tissues. You can monitor the efficiency of RNA recovery by adding an equal amount of a ^{35}S-labelled control RNA, synthesized *in vitro*, to each cell or tissue sample at the start of the RNA isolation procedure.

If you opt to load equal amounts of total (or polyadenylated) RNA in each track, do not assume that two RNA samples giving identical readings on a spectrophotometer contain identical amounts of intact mRNA. If you do, you may encounter one of the following problems:

- the samples may be contaminated with different amounts of genomic DNA,
- the RNA in one sample may be partially degraded,
- if your samples consist of oligo(dT)-selected polyadenylated RNA, the degree of enrichment for polyadenylated RNA may differ in the two samples.

◇ Put another way, the polyadenylated RNA samples may be contaminated with different amounts of ribosomal RNA, transfer RNA, and unprocessed RNAs.

How can you avoid these problems? Always check the quality of your RNA before use by running an aliquot on an agarose gel containing ethidium bromide. This will tell you whether the RNA is intact and whether the samples contain roughly equal amounts of RNA (*Figure 4.4*). However, you can never be absolutely sure that what looked good one day will also look good the next; one of your samples may have degraded in the interim. A better approach, therefore, is to examine the quality of the electrophoresed RNA on the gel that you plan to blot, by staining the electrophoresed gel with ethidium bromide (see section 4.4). For reasons discussed in section 4.4, however, you may not wish to do this. The best option, and one that many referees insist on, is to probe your blot for the RNA you are studying and then to strip off the probe and reprobe the blot for an RNA species that is known to be present at the same level in each of the cell types you are analysing. If this second probe gives similar autoradiographic signals in each track then you can be confident that you have loaded similar levels of intact mRNA (*Figure 4.5*). This begs the question of which control probe to use. Many workers use a probe for β-actin mRNA and assume that levels of this mRNA remain constant in the cell types that they are using. This is often a correct assumption, but there are exceptions. Other options are to reprobe for β-tubulin or glucose-6-phosphate dehydrogenase mRNAs. The

◇ One of us found to his cost that β-actin mRNA levels change when promyelocytic HL60 cells are induced to differentiate.

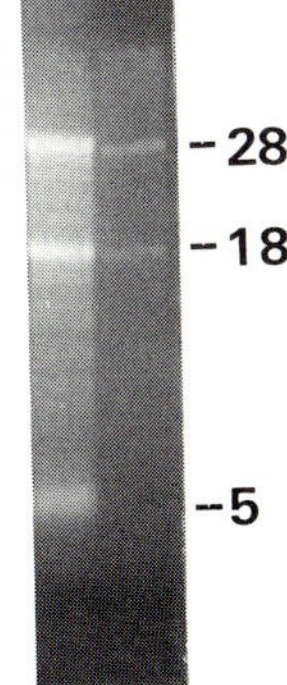

Fig 4.4

Samples containing different concentrations of rat kidney total RNA were electrophoresed on a 1 % agarose-MOPS-formaldehyde gel and stained with ethidium bromide. The 28S, 18S, and 5S rRNA bands are indicated

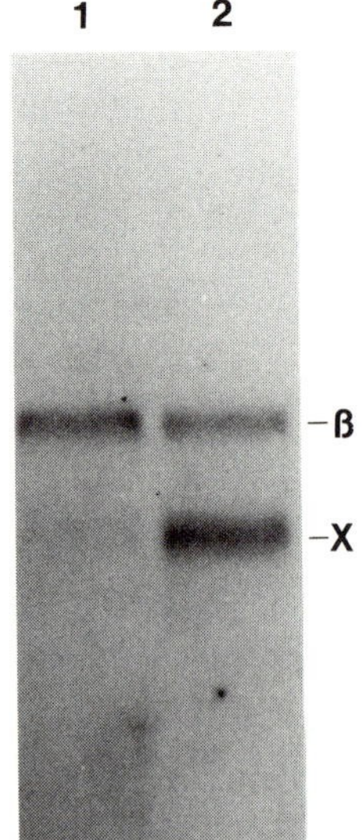

Fig 4.5

The cDNA clone 'X' was isolated by a differential library screening protocol designed to identify mRNAs present at higher levels in Epstein–Barr virus (EBV)-infected BL cells than in uninfected cells (Brickell and Patel 1988). Poly(A)-enriched mRNA from uninfected (1) and EBV-infected (2) BL cells was electrophoresed on an agarose gel, blotted on to nylon and probed with radiolabelled clone X. The probe detected an 800-nucleotide m*RNA* (X). The membrane was re-hybridized with a probe for 1.8 kb β-actin mRNA (β). Given that β-actin mRNA levels are equal in both cell types, the result shows that X mRNA is indeed more abundant in EBV-infected BL cells than in uninfected cells.

important thing is to make sure that the control probe is suitable for your system.

3.2. Quantitation

Having convinced yourself that a difference in the intensity of an autoradiographic signal in two different tracks really does represent differential accumulation of mRNA in two different cell types, can you quantitate this? The answer is yes, but with great care. First, you must assume that the amount of radioactive probe that has bound is proportional to the amount of the specific mRNA present. Secondly, you must find a way of measuring the amount of radioactive probe that has bound. There are three ways you could do this:

◇ This is probably a safe assumption in most northern blotting experiments.

- scintillation counting of excised bands,
- scanning densitometry,
- phosphorimagery.

3.2.1 Scintillation counting of excised bands

◇ You cannot re-use the membrane, of course.

Match the membrane to the autoradiograph, cut out (with a scalpel blade) the parts of the membrane that correspond to autoradiographic signals, place them in scintillation fluid and use a scintillation counter to measure the amount of radioactive probe that has bound to each piece of membrane. This will give you a fairly reliable estimate of the relative levels of RNA in the different bands.

3.2.2 Scanning densitometry

Measure the intensities of the autoradiographic signals using a scanning densitometer. This has an air of respectability, but suffers from some serious problems. The main drawback is that the response of X-ray film to a radioactive signal is linear over only a rather limited range (see Chapter 5, *Figure 5.5*). You can only accurately estimate the relative levels of probe bound in two tracks if the autoradiographic signal in both tracks lies within the linear range of the film. This creates two problems:

- What is the linear range of the film? You can only answer this question rigorously (rather than by guessing) if you calibrate the film. You could do this in advance, perhaps by radiolabelling a DNA fragment, adding a range of dilutions into a series of wells in an agarose gel, electrophoresing the samples, exposing the gel to X-ray film for a defined period of time, and, finally, measuring the intensity of the autoradiographic signal in each track by scanning densitometry. By plotting a graph of densitometer reading against DNA dilution you could determine the range of densitometer readings that lie within the film's linear range for the exposure time used. You would have to obtain such data for a range of different exposure times and then choose a suitable exposure time for your northern hybridization experiment.
- The second problem is that if you want to compare the intensities of the autoradiographic signals in two tracks of a northern blot (carrying two different RNA samples), but find that the intensities are very different, it may be difficult to find an exposure time at which both signals lie within the linear range of the film. In this case, you should blot and hybridize a gel on which you have loaded a series of dilutions of each sample. If you then obtain densitometer readings for the two dilution series, you should be able to determine the linear range of the film, choose readings within that range, and multiply by the corresponding dilution factor to obtain an estimate of the relative abundance of your mRNA in the two samples.

◇ Such an experiment could rapidly become very unwieldy if you want to compare more than two samples.

3.2.3 Phosphorimagery

If this all sounds like a lot of hard work, for dubious returns, save up as we are doing, and buy a phosphorimager. This wonderful machine will scan the membrane and produce a computer display that resembles an autoradiograph. You can then use the computer's software to tell you the amount of radioactivity that gives rise to each band.

It should be clear from the above that quantifying the relative abundance of an mRNA species in several cell or tissue types is possible, but fraught with difficulty. If the differences in abundance are large (more than 5- or 10-fold) you can interpret your data with some confidence. If the differences are small, you should be more circumspect.

4. Gel electrophoresis of RNA samples

We have discussed the problems that can arise in interpreting northern blots and have seen that these influence the way in which you will perform your experiment. How, then, do you do a northern blot?

4.1 Gel systems

As with DNA, RNA molecules of different sizes can be separated by electrophoresis through agarose or polyacrylamide gels. We will confine this discussion to the use of agarose gels, which are by far the most widely used means of analysing RNA populations. There are two systems in general use for performing agarose gel electrophoresis of RNA samples under denaturing conditions. In the first, referred to as the *formaldehyde gel*, both the sample loading buffer and the agarose gel itself contain formaldehyde as a denaturing agent (Lehrach *et al.* 1977). In the second, referred to as the *glyoxal gel*, RNA samples are denatured in a mixture of glyoxal and dimethyl sulphoxide (DMSO) before loading (Thomas 1980).

◇ The gel itself does not contain glyoxal.

Your first task is to decide which gel electrophoresis system to use. Most people use formaldehyde gels because they are easier to run than glyoxal gels and have equally good resolving power. However, the bands of RNA detected following northern blotting are usually sharper when the RNA has been electrophoresed on a glyoxal gel and this may be important under some circumstances.

4.2 Formaldehyde gels

Formaldehyde denatures RNA by reacting covalently with the amine groups of adenine, guanine, and cytosine and so preventing the formation of G-C and A-T base pairs. Clearly, formaldehyde will also prevent hybridization of a probe to the RNA on the membrane after blotting. The reaction of formaldehyde with RNA is therefore reversed after blotting, before hybridization.

◇ Formaldehyde has the structure:

```
      O
      ||
  H—C—CH3
```

Since formaldehyde cross-links to proteins, it will also help to inactivate RNases in the gel.

4.2.1 Preparing formaldehyde gels

The composition of a formaldehyde gel differs from that of a standard agarose gel used for electrophoresis of DNA. However, the same general principles of gel preparation should be followed (see Chapter 2, section 3). To make a formaldehyde gel you will need MEA buffer, formaldehyde, good quality de-ionized water, and agarose.

◇ Use good quality agarose, as discussed in Chapter 2, section 3.1).

◇ 10 × MEA buffer contains 200 mM MOPS (sodium salt), 50 mM sodium acetate pH 7.0, 10 mM EDTA.

MEA buffer contains 3-(N-morpholino) propanesulphonic acid (MOPS), EDTA, and sodium acetate. The MOPS and sodium acetate are there to buffer the gel. MOPS is used in preference to Tris because formaldehyde would react with the amine groups of Tris. The EDTA is there to chelate divalent cations and so inhibit nucleases.

MEA can be prepared in advance as a stock solution at 5 or 10 times the working concentration. Most laboratory manuals tell you to sterilize MEA by filtration because MOPS breaks down when autoclaved. However, this is very tedious if you have a large volume of stock solution to sterilize and, in fact, there are no real problems in autoclaving MEA, using a standard 20 minute cycle. If MEA is stored for long periods in the light the MOPS does break down. The MEA becomes progressively more yellow and there will come a time

◇ MEA becomes slightly straw-coloured after autoclaving. This will not harm your experiment.

when it no longer buffers well. We therefore routinely store MEA out of the light, in a refrigerator.

◇ Formaldehyde fumes are both toxic and unpleasant.

◇ You can remove formic acid by de-ionizing the solution, but it is an unpleasant business.

Use Analar grade formaldehyde, supplied as a 37 per cent (v/v) solution (12.3 M) in water, straight from the bottle. Formaldehyde can be oxidized to formic acid, which will destroy RNA by acid hydrolysis. If the pH of the formaldehyde is lower than 4.0, do not use it. Formaldehyde is a volatile chemical and the fumes are toxic and highly irritating. It must be used in a fume hood.

◇ If you add formaldehyde and MEA at a higher temperature you will generate a lot of formaldehyde fumes and will further degrade the MOPS.

Having assembled the components of the gel, dissolve the agarose by boiling in an appropriate volume of water (see Chapter 2, section 3.3). For analysis of RNA up to approximately 3 kb in length, we routinely prepare gels with a final agarose concentration of 1.5 per cent (w/v). For analysis of larger RNAs, a 1 per cent (w/v) agarose gel would be better. Transfer the agarose solution to a fume hood and, when it has cooled to 60 °C, add the appropriate quantities of MEA and formaldehyde. Quickly but gently mix the gel solution, taking care not to make bubbles, and pour it into the gel apparatus as discussed in Chapter 2, section 3.3. Pour the gel, in the fume hood. When the gel has been allowed to set for at least 30 minutes, place it in the gel electrophoresis apparatus and cover it with running buffer. Connect the power pack so that it can be switched on immediately it is needed. If possible, you should run the gel in a fume hood since the warming of the gel during electrophoresis will generate formaldehyde fumes. These are not generally excessive, however, and the gel can be run on the open bench provided that you do not lean too closely over the apparatus or breath too deeply when you remove the lid of the gel apparatus after electrophoresis. You should, of course, put a hazard label on the apparatus so that your colleagues do not inhale it by accident.

4.2.2 Preparing RNA samples for formaldehyde gels

◇ There is nothing worse than searching for a power pack or indulging in complicated calculations of RNA concentrations whilst your thawed RNA samples degrade.

It is outside the scope of this book to discuss methods for isolating RNA from cells. Such techniques are discussed by Wilkinson (1991). We assume that your −70 °C freezer contains good quality RNA samples of known concentration, in water. As discussed above, RNA is degraded by RNases and it is not a good idea to leave it unfrozen in any form for longer than is absolutely necessary. Leave the RNA in the freezer until the last possible moment. Before you thaw the RNA, you should pour the gel, set up the gel electrophoresis apparatus, set up the power pack, prepare and dispense the sample loading buffer, and calculate exactly how much RNA to add to each well.

◇ The formamide disrupts hydrogen bonding and so allows the formaldehyde to react with the bases.

The sample loading buffer that we use contains MEA, formaldehyde, and formamide. These three components should be mixed in the ratio 1:1.8:5 (MEA:formaldehyde:formamide) on the day of use. MEA and formaldehyde can be taken from the same stocks used to prepare the gel. Formamide frequently contains ionic contaminants, such as ammonium formate, and some of these can hydrolyze RNA. It is therefore essential that you remove them by de-ionization. To do this, add a mixed-bed ion-exchange resin, such as 20 per cent (w/v)

Dowex XG8, stir for an hour, and filter twice through Whatman No. 1 paper. Store aliquots of the de-ionized formamide at –20 °C.

If 3 volumes of this sample loading buffer are added to 1 volume of RNA sample in water, the mixture is dense enough to be loaded easily into the well of a submerged gel. In addition, the refractive properties of the mixture enable you to see that it has entered the well properly, although it does take some practice to judge this. Some people include glycerol in the sample loading buffer to make the mixture more dense and so easier to load. You can also add bromophenol blue and xylene cyanol FF to the sample loading buffer, enabling you to judge more easily whether all of the sample has entered the well properly. We omit glycerol and marker dyes from our sample loading buffer on the principle that the fewer things that are in it, the less harm can come to our RNA. If you omit marker dye from the sample loading buffer, you must remember to add some to a spare well at the edge of the gel, since you can only judge when to stop electrophoresis by watching the migration of the marker dye.

◇ It may be that we are a little paranoid in this respect.

4.2.3 How much RNA should be loaded?

In deciding how much RNA to load into each well of the gel you should bear the following points in mind:

- Before loading, each RNA sample will be mixed with three times its volume of sample loading buffer. The final volume of the sample must not exceed the capacity of the well. Check the maximum volume that the wells will accommodate. This will typically be between 25 μl and 50 μl, depending on the design of the apparatus.
- Do not load more than about 20 μg of RNA into a standard 3 × 1 mm well, or the resolving power of the gel and the sharpness of the RNA bands that you detect will deteriorate.
- In general, the aim of a northern blotting experiment is to detect a specific mRNA. Note that mRNA constitutes only a very small proportion of total cellular RNA (*Table 4.1*).
- Estimate the likely abundance of the specific mRNA to be detected. If it is relatively abundant (as a rough guide, if it constitutes more than 0.1 per cent of all the mRNAs in the cell), then it will usually be sufficient to load 10–20 μg of total RNA per well. To detect less abundant mRNAs you would need to load more than 20 μg of total RNA per well. As noted above, this is not a

◇ We get tighter bands if we add this amount of RNA to a wider well, or two 3 × 1 mm wells taped together.

◇ Polyadenylated mRNA typically constitutes approximately 50 per cent of the RNA in such enriched samples.

Table 4.1 Typical abundance of different types of RNA in a eukaryotic cell

RNA type	Abundance
Nuclear mRNA precursors	6%
Nuclear rRNA precursors	4%
Nuclear tRNA precursors and small nuclear RNAs	1%
Cytoplasmic ribosomal RNA (rRNA)	71%
Cytoplasmic transfer RNA (tRNA)	15%
Cytoplasmic messenger RNA (mRNA)	3%

good idea and so you should use RNA that has been enriched for polyadenylated mRNA by affinity chromatography using oligo(dT). Whilst you could add up to 20 µg of such RNA per well, 1–5 µg is usually enough. These considerations are rather vague and in the end you may have to determine empirically the amount of RNA that you should add. Thus, if you cannot detect the mRNA that interests you using 20 µg of total RNA, you should try using samples enriched for polyadenylated mRNA. On the other hand, if you obtain a very intense signal using 20 µg of total RNA, you will be able to reduce the quantities used in future experiments.

- Finally, remember to load a range of dilutions of each sample if you intend to quantitate autoradiographic signals by scanning densitometry (see section 3.2.2).

At the end of these deliberations you should have decided what volume of each RNA sample to take and what volume of sample loading buffer to add (three times the volume of the sample). The final volume should be less than the maximum capacity of the well.

◇ If you do not thaw the sample completely, the RNA concentration in the thawed portion will differ from that in the frozen portion and, in consequence, from that in the completely thawed solution (which you used to determine the RNA concentration).

◇ Replace the stock RNA samples in the freezer as soon as you have finished with them.

◇ Others suggest 95 °C for 2 minutes. The heat, like the formamide, disrupts hydrogen bonds and allows the formaldehyde to react with the bases.

Label the correct number of clean snap-cap tubes and dispense the required amount of sample loading buffer into each of them. Remember to include at least one tube for molecular weight markers, if you are going to use them (see section 4.4). Now you may thaw your RNA samples! Thaw them rapidly and completely, mix the contents gently, and place them on ice. Dispense the required amount of each sample to the appropriate tube, mixing the sample with the sample loading buffer as you do so. Before loading the dispensed RNA on to the gel, the RNA must be thoroughly denatured by incubation at 65 °C for 15 minutes. During this period, some people pre-run the agarose gel for 5 minutes at 5 volts/cm (see Chapter 2, section 3.6). In some old protocols, denaturation of RNA before electrophoresis was performed in the presence of methylmercuric hydroxide (Bailey and Davidson 1976). This is highly toxic and we do not recommend that you revive this ancient practice.

4.2.4 Running formaldehyde gels

◇ Formaldehyde fumes will have accumulated under the lid, so do not get too close, or breath too deeply, when you remove it.

◇ Sodium hydroxide treatment is only worth considering if you have difficulty detecting large RNAs.

Formaldehyde gels should be run at 3–4 volts/cm, taking account of the general points made about agarose gel electrophoresis in Chapter 2, section 3.6. At the end of the run, judged by the progress of the marker dye, carefully remove the gel from the apparatus. Formaldehyde gels are more fragile than non-denaturing agarose gels, so handle them with care. For this reason, and in order to minimize diffusion of the RNA in the gel and degradation of the RNA by RN ases, you should limit the number of manipulations of the gel prior to blotting. Some protocols suggest that you wash the gel in DEPC-treated water to remove formaldehyde and incubate the gel in sodium hydroxide solution to partially hydrolyze the RNA and so facilitate its transfer to the membrane. We suggest that you omit all of these steps and blot the gel immediately it is removed from the gel apparatus.

4.3 Glyoxal gels

◇ Glyoxal has the structure:

```
        O
      /   \
 HO—C       C—OH
    |       |
 HO—C       C—OH
      \   /
        O
```

Glyoxal denatures RNA by binding covalently to guanine residues and so preventing the formation of G-C base pairs. Like formaldehyde, glyoxal will prevent hybridization of a probe to the RNA on the membrane after blotting. The reaction of glyoxal with guanine must therefore be reversed after electrophoresis, either before or after blotting.

4.3.1 Preparing glyoxal gels

◇ The interaction between glyoxal and guanine is unstable at high pH.

◇ The concentration of agarose should be 1–1.5 per cent (w/v), as discussed in section 4.2.1.

◇ Sodium iodoacetate is toxic. Handle it with care.

Glyoxal gels do not themselves contain glyoxal. Rather, the glyoxal is present in the RNA samples loaded on to the gel. Glyoxal gels are prepared simply by dissolving agarose in sodium phosphate buffer by boiling (see Chapter 2, section 3.1). After cooling the gel solution to 70 °C, add solid sodium iodoacetate to a final concentration of 10 mM. The purpose of the sodium iodoacetate is to inactivate RNase A, which it does by interfering with disulphide bonds in the protein. Sodium iodoacetate is not required in formaldehyde gels, since the formaldehyde will inactivate RNases.

The gel mixture may be poured into the gel apparatus once it has cooled to 50 °C, as before.

4.3.2 Preparing RNA samples for glyoxal gels

As before, do not thaw your RNA stocks until you need them. Before loading on to a glyoxal gel, RNA is denatured by incubation for 1 hour at 50 °C in the presence of glyoxal and DMSO. The DMSO and high temperature disrupt intramolecular hydrogen bonds, allowing the glyoxal to react with guanine residues. Sodium phosphate is included in the mixture as a buffer. Note that:

◇ Glyoxal solutions contain inhibitors of polymerization, but these do not affect the RNA.

◇ Glyoxylic acid will be ionized in solution and will bind to the resin. Glyoxal, being non-ionic, will stay in solution.

- Glyoxal is usually obtained as a 40 per cent (6.89 M) aqueous solution. Glyoxal is readily oxidized in air to yield glyoxylic acid. This hydrolyzes RNA and so must be removed before use. This can be done by de-ionizing the solution. To do this, pass it through a mixed-bed resin, such as Bio-Rad AG 501-X8, until its pH is greater than 5.0. Store de-ionized glyoxal in small aliquots in tightly capped tubes at –20 °C. Once a tube is re-opened, its contents should be used or discarded.
- Good quality DMSO may be used straight from the supplier's bottle.
- Most manuals recommend that stock sodium phosphate buffer should be treated with DEPC to inactivate RNases and then sterilized by autoclaving.

After incubation, chill the RNA samples on ice. You may have to centrifuge the tubes briefly since some of the solution will have evaporated during incubation and condensed at the top of each tube. Mix the samples with glyoxal gel loading buffer (which contains glycerol, bromophenol blue and xylene cyanol in a sodium phosphate buffer) and load them on to the gel immediately.

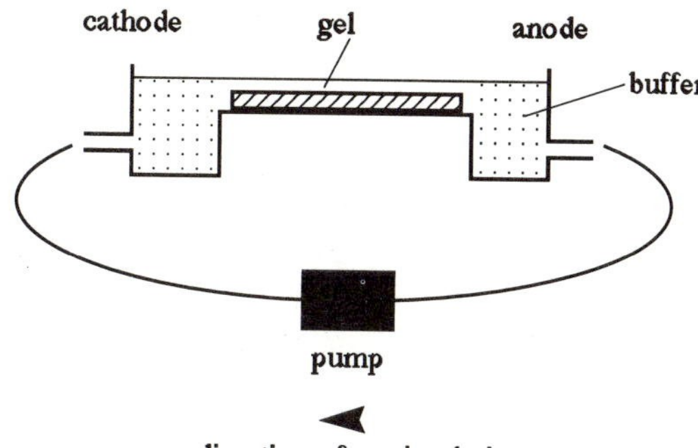

Fig 4.6

Cross-section of gel electrophoresis tank, illustrating re-circulation of buffer during agarose gel electrophoresis

4.3.3 Running glyoxal gels

Glyoxal gels are run submerged in sodium phosphate buffer at 3–4 volts/cm. It is very important that a large pH gradient does not build up across the gel during electrophoresis, because glyoxal will dissociate from RNA if the pH rises much above 8.0. An even pH can be maintained by recirculating the buffer between the reservoirs at each end of the gel (*Figure 4.6*). A technically simpler, but more laborious, solution to this problem is to replace the sodium phosphate gel running buffer with fresh buffer every 30 minutes during electrophoresis.

After electrophoresis, glyoxal must be removed from the RNA. If you intend to blot on to a nylon membrane, you can removal glyoxal after blotting by soaking the membrane in 20 mM Tris-HCl (pH 8.0) or in 50 mM sodium hydroxide. Alternatively, you can treat the gel with sodium hydroxide before blotting. The possible advantages of the latter are that:

- sodium hydroxide will also partially hydrolyze the RNA and so increase the efficiency of transfer of large RNAs to the membrane
- if blotting is performed under alkaline conditions, RNA will bind irreversibly to the nylon membrane without the need for further treatment (see section 5).

The disadvantage of treating the gel before blotting is that there is increased opportunity for diffusion of RNA in the gel, which can give rise to fuzzy bands.

◇ Nitrocellulose filters become very brittle during sodium hydroxide treatment, which should be avoided.

If you are using a nitrocellulose filter, you should remove the glyoxal after blotting, by soaking the filter in 20 mM Tris-HCl (pH 8.0).

4.4 Molecular size markers

◇ RNA molecules migrate faster than DNA molecules of the same length in formaldehyde gels.

◇ To glyoxylate heat-denatured DNA, treat it in the same way as RNA.

DNA fragments of known length are not suitable for use as markers on formaldehyde gels, since DNA and RNA molecules of the same length migrate at different rates. Glyoxylated, single-stranded DNA fragments of known length can be used in glyoxal gels, however, because they do migrate at the same rate as glyoxylated RNA molecules of the same length. There are other options:

- use 28S and 18S rRNA
- use other RNAs of known length.

4.4.1 28S and 18S rRNA size markers

◇ The quoted length of human 28S rRNA varies between 4.2 kb and 6.3 kb. Many lab manuals heartily recommend 28S and 18S rRNA as size markers but do not state their length.

◇ The sizes of 28S and 18S rRNAs differ between organisms. For example, in the chicken, they are 4.6 kb and 1.8 kb long, respectively (Darnell *et al.* 1990).

The simplest option is to use the 28S rRNA and 18S rRNA in your RNA samples as markers (*Figure 4.3*). Everybody agrees that this is a good idea, but no two textbooks or lab manuals agree on the lengths of human 28S rRNA and 18S rRNA. Darnell *et al.* (1990) state that they are 5.1 kb and 1.9 kb long, respectively, and we use these values. If you are working with plant RNA, the chloroplast rRNAs (23S and 16S) may also be visible. Bacteria also have rRNAs of 23S and 16S (2904 and 1541 nucleotides in *Escherichia coli*), but in some species one or both molecules are present as two or more fragments.

◇ If you are using tiny quantities of RNA, or want to detect a rare RNA, it may be best not to put ethidium bromide in the gel.

◇ To avoid cutting away part of a track containing an important sample, leave at least one vacant track between the marker and the samples.

◇ Stain for 5–10 minutes in 20 × SSC containing 0.5 μg/ml ethidium bromide. 20 × SSC contains 3 mM sodium chloride and 300 mM sodium citrate.

In formaldehyde gels, rRNAs can be visualized by including ethidium bromide in the gel during the run and then viewing the gel with UV radiation (see Chapter 2, section 3.7). If there is enough RNA in each track, it may be possible to see the stained rRNA bands on the membrane after blotting, in normal light. However, ethidium bromide can reduce the efficiency of hybridization to RNA subsequently blotted from the gel. It may therefore be best to load a marker sample in an outside track of the gel and to cut off this track after electrophoresis, stain it with ethidium bromide, and view it with UV radiation.

Do not put ethidium bromide in glyoxal gels during electrophoresis because it reacts with glyoxal and interferes with its ability to denature the RNA. Stain the RNA after electrophoresis by soaking the entire gel, or a track cut from it, for 5–10 minutes in 10 mM sodium phosphate (pH 7.0) containing 0.5 μg/ml ethidium bromide. The rRNA bands can then be seen in UV radiation in the normal way.

If you wish to avoid staining the gel altogether, it is possible to stain RNA with methylene blue after it has been transferred to a membrane and even after hybridization.

The use of 28S and 18S rRNAs as molecular size markers has the important disadvantage that only two data points are obtained for a standard curve. You will therefore be able to obtain only a very approximate estimate of the size of the RNA species in which you are interested. If this species is larger than 28S rRNA or smaller than 18S rRNA, the accuracy of your estimate will be particularly poor.

4.4.2 Other RNA size markers

A better, but more expensive, alternative to the use of 28S and 18S rRNA as size markers is to purchase one of the mixtures of RNAs of known size that are now commercially available. For example, Gibco-BRL market a '0.16–1.77 kb RNA ladder' (catalogue number 520-5623SA) and a '0.24–9.5 kb RNA ladder' (catalogue number 520-5620SA). Alternatively, you could make your own mixture by using T3 or T7 RNA polymerase to transcribe RNAs *in vitro* from a series of DNA fragments cloned in a suitable vector, such as pBluescript II (Alting-Mees *et al.*, 1992).

5. Blotting the gel

◇ The use of higher concentrations of sodium hydroxide will result in extensive hydrolysis of the RNA.

Formaldehyde and glyoxal gels are blotted in the same way as non-denaturing gels (see Chapter 3). The only difference is that if you are blotting under alkaline conditions, you should use 7.5 mM sodium hydroxide for transfer of RNA. However, we do ***not*** recommend that you perform northern blotting under alkaline conditions. In addition to the problems discussed in Chapter 3, section 3, in the context of Southern blotting, you risk hydrolyzing the RNA.

RNA can be transferred to nitrocellulose filters or nylon membranes and all of the advantages of nylon membranes, discussed in Chapter 7, apply. After blotting, membranes carrying RNA should be rinsed in the same way as membranes carrying DNA (Chapter 3, section 1.3). Methods for fixing RNA to membranes are also the same as for DNA. However, in addition, if you used a formaldehyde gel, you should bake the membrane at 80 °C for 2 hours to remove residual formaldehyde from the RNA. If you used a glyoxal gel and did not remove the glyoxal before blotting, remember to do this now (see section 4.3.3).

◇ Failure to reverse the formaldehyde and glyoxal reactions will seriously impair the efficiency of hybridization.

6. Further reading

Perbal, B. (1988). *A practical guide to molecular cloning*, (2nd edn), pp. 526–42, Wiley, New York.

Sambrook, J., Fritsch, E.F., and Maniatis, T. (1989). *Molecular cloning: a laboratory manual*, (2nd edn), Vol. 1, pp. 7.37–7.52, Cold Spring Harbor Laboratory Press.

Wahl, G.M., Meinkoth, J.L., and Kimmel, A.R. (1987). Northern and Southern blots. *Methods in Enzymology*, **152**, 572–81.

7. References

Alting-Mees, M.A., Sorge, J.A., and Short, J.M. (1992). pBluescript II: multifunctional cloning and mapping vectors. *Methods in Enzymology*, **216**, 483–95.

Alwine, J.C., Kemp, D.J., and Stark, G.R. (1977). Method for detection of specific RNAs in agarose gels by transfer to diazobenzyloxymethyl-paper and hybridization with DNA probes. *Proceedings of the National Academy of Sciences, USA*, **74**, 5350–4.

Aviv, H. and Leder, P. (1972). Purification of biologically active globin messenger RNA by chromatography on oligothymidylic acid-cellulose. *Proceedings of the National Academy of Sciences, USA*, **69**, 1408–12.

Bailey, J.M. and Davidson, N. (1976). Methylmercury as a reversible denaturing agent for agarose gel electrophoresis. *Analytical Biochemistry*, **70**, 75–85.

Brickell, P.M. and Patel, M. (1988). Structure and expression of c-*fgr* protooncogene mRNA in Epstein-Barr virus converted cell lines. *British Journal of Cancer*, **58**, 704–9.

Darnell, J., Lodish, H. and Baltimore, D. (1990). *Molecular cell biology*, (2nd edn), p. 271, Scientific American Books, New York.

Lehrach, H., Diamond, D., Wozney, J.M., and Boedtker, H. (1977). RNA molecular weight determination by gel electrophoresis under denaturing conditions: a critical re-examination. *Biochemistry*, **16**, 4743–51.

Marzluff, W.F. and Huang, R.C.C. (1985). In *Transcription and translation: A practical approach*, (ed. B.D. Hames and S.J. Higgins), pp. 89–129. IRL Press at Oxford University Press.

Thomas, P.S. (1980). Hybridization of denatured RNA and small DNA fragments transferred to nitrocellulose. *Proceedings of the National Academy of Sciences, USA*, **77**, 5201–5.

Wilkinson, M. (1991). Purification of mRNA. In *Essential molecular biology: A practical approach*, (ed. T.A. Brown), pp. 69–87. Vol. I, IRL Press at Oxford University Press.

5 Dot and slot blotting

1. What are dot blots and slot blots?

Sometimes you may wish to bind a DNA or RNA sample to a membrane without first separating its components by electrophoresis. This can be done by pipetting DNA or RNA on to a membrane as a small spot (Kafatos *et al.* 1979). This is not ideal because the samples tend to spread out over different distances depending upon how quickly they are applied. Spreading can be minimized if each sample is applied in a series of 2 µl aliquots, with each aliquot being allowed to dry before addition of the next. However, it is still difficult to obtain tight spots of reproducible shape and size.

To make this process more reproducible the *dot blot manifold* was born. Manifold designs are themselves manifold. A typical model, illustrated in *Figures 5.1* and *5.2A*, consists of three perspex blocks that can be clamped together with screws and nuts. The upper block has a regular array of circular holes to which samples can be added, the middle block has a matching array of holes, and the lower block has a cavity with an outlet that can be connected to a vacuum pump. The blocks are clamped together with a nylon membrane or nitrocellulose filter sandwiched between the upper and middle blocks. Samples of DNA or RNA in solution are added to the holes at the top and a vacuum applied to the outlet at the bottom. The vacuum sucks the solution downwards, leaving the DNA or RNA bound to the membrane. The advantage of these systems is that samples are confined to strictly defined areas of the membrane, making it easier to compare the intensities of hybridization signals corresponding to different samples. An example of a result obtained by dot blotting is shown in *Figure 5.3A*.

◇ Other models consist of just two blocks (*Figure 5.2B*), but work to the same principle.

The second generation of blotting manifold was the *slot blot manifold*. This is essentially identical to the dot blot manifold except that

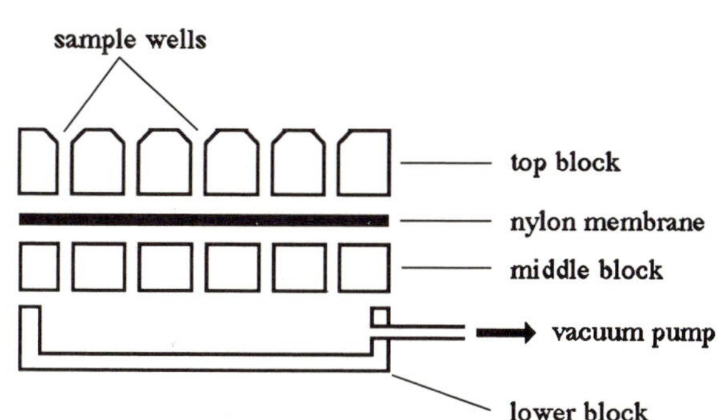

Fig 5.1

Cross-section through a typical dot/slot blot manifold. The blocks are held together by screws and nuts (not shown)

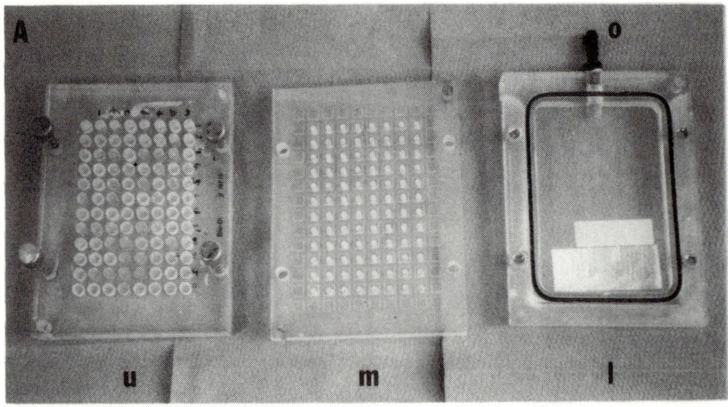

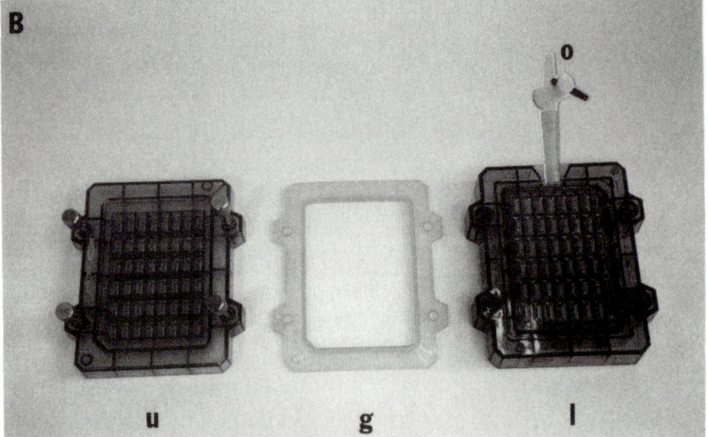

Fig 5.2

Two types of dot/slot blot manifold. **A**. A dot blot manifold comprising 3 perspex blocks (u, upper; m, middle; l, lower), as described in the text. **B**. A slot blot manifold comprising 2 plastic blocks (u, upper; l, lower) and a rubber gasket (g), which is placed between the blocks. In both cases, the lower block has an outlet (o) to a vacuum pump

the sample wells are in the form of a narrow slot, rather than of a circle (*Figures 5.2B* and *5.3B*). There were two main reasons for this modification:

- Slot-shaped hybridization signals are easier to scan with a densitometer. In particular, it is easier to perform multiple independent scans of a single slot than of a single dot.
- Some hybridization artefacts are more easily detectable with slot-shaped wells. For example, the presence of contaminating protein, or poor washing of the membrane after hybridization, can cause a non-specific background spot of radioactivity to appear on the autoradiograph (*Figure 5.4A*). With dot blot manifolds, you may confuse this with a real signal. In contrast, it is unlikely that an artefactual signal would extend over the entire length of a slot in a slot blot manifold.

◇ When such spots are seen on autoradiographs of Southern and northern blots, they are immediatly apparent as artefacts.

1.1 Why do a DNA dot/slot blot?

Suppose, for example, you had a panel of cDNA clones corresponding to a number of genes that you were interested in, and you wanted to know which of these genes were expressed in a particular cell type. One way to address this question would be to perform a series of

A

1 2

3 4

B

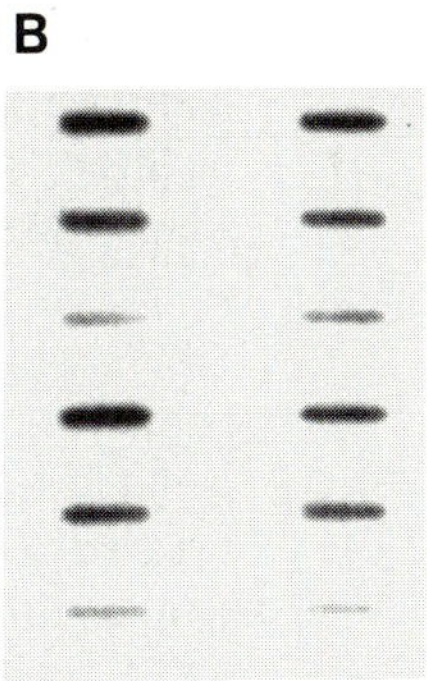

Fig 5.3

A. Autoradiograph of a dot blot. Probe hybridized strongly to sample 2, weakly to sample 3 and very weakly to sample 1. There was no detectable hybridization to sample 4. The signal for example 2 lies outside the film's linear range. The signal for sample 3 lies within the linear range, but is uneven, being most intense around the rim of the dot. It would be hard to quantify by scanning densitometry. B. Autoradiograph of a slot blot. These signals would be much easier to quantify than those in A

A

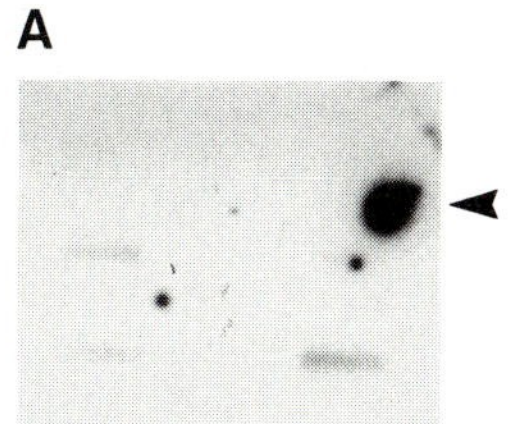

B

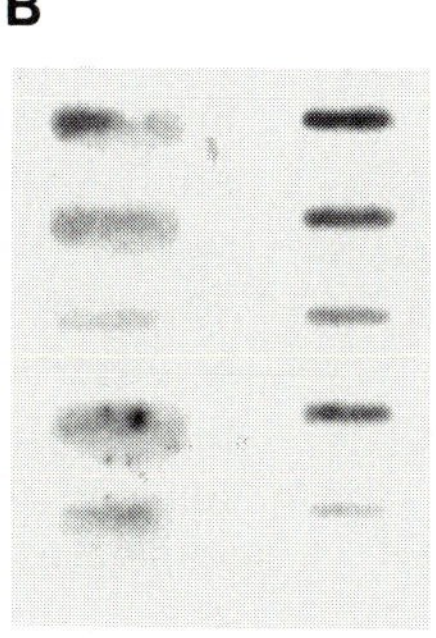

Fig 5.4

Some things that can go wrong with dot/slot blots. A. Autoradiograph of a slot blot with an obviously artefactual signal (arrowed). This could be mistaken for a real signal if it appeared on a dot blot at a position where a dot might be expected. B. Slots in the left-hand column have leaked, allowing samples to spread

northern blots and probe each with a different member of the cDNA panel.

However, a more rapid approach would be to bind all of the cloned cDNAs to a single membrane, in an ordered array, and to probe this membrane with radiolabelled mRNA prepared from the cell type under study. Those cDNAs that gave hybridization signals would represent genes that were transcribed in the cell type from which you obtained the mRNA.

◇ Instead of radiolabelling the mRNA, you could synthesize radiolabelled cDNA, using the mRNA as a template.

You can also use the dot/slot blot method to compare the levels of particular mRNAs in two cell types. In this case, you would need to make two membranes carrying an identical panel of cloned cDNAs, and would then probe these with radiolabelled mRNA from each of the cell types. In such an experiment, differences in the intensity of hybridization of the two probes to a particular cDNA clone would reflect differences in the abundance of the corresponding mRNA in the cell types from which the two probes were prepared.

◇ This is determined by a balance between the rates of synthesis and degradation of the mRNA in the two cell types.

As with northern blotting, such an experiment allows you to compare mRNA levels in two cell types but does *not* allow you to draw any conclusions about transcription rates (Chapter 4, section 2). However, the duplicate membranes could also be used in a *nuclear run-on assay* (Marzluff and Huang 1985; Lillycrop *et al.* 1992), which does allow you to compare transcription rates. In this case, you would prepare the probes not by radiolabelling mRNA isolated from cells, but by isolating cell nuclei and allowing them to transcribe their genes in the presence of radiolabelled uridine 5′-triphosphate (UTP). In such probes, the abundance of a particular RNA species reflects the efficiency of its synthesis by transcription and is unaffected by the stability of the corresponding mRNA. In this case, therefore, differences in the intensity of hybridization of the two probes to a particular cDNA clone *would* reflect differences in the rate at which the corresponding gene is transcribed in the two cell types.

1.2 Why do an RNA dot/slot blot?

RNA dot/slot blots provide a rapid means of analysing a large number of RNA samples for the presence of a particular mRNA species. For example, you could blot a large number of RNA samples on to a membrane and then hybridize them with a radiolabelled probe for an mRNA of interest. The intensity of hybridization to each RNA sample would reflect the abundance of the mRNA in each tissue type. This method is much quicker than northern blotting because there is no electrophoresis step, and it enables you to deal with more samples than you could accomodate on a northern blot. You would have to run and blot a large number of gels to accomodate the number of samples that can be handled in a single dot/slot blot.

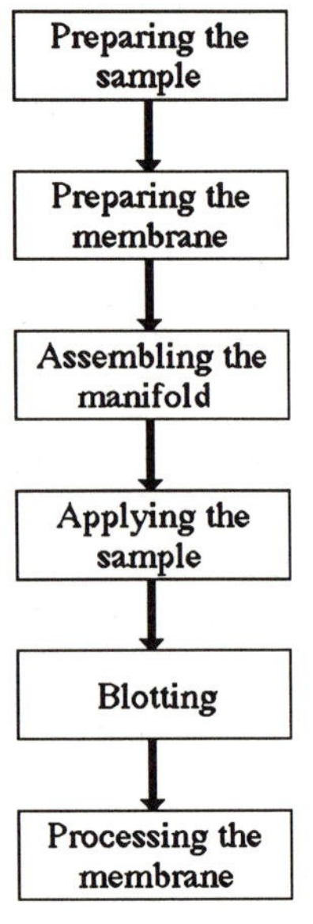

2. How to perform a dot/slot blot

There are many different protocols for dot/slot blotting. Manufacturers of blotting manifolds and nylon membranes distribute advice on how best to use their products, and you should follow this carefully. There are six steps to consider:

(1) Preparation of the sample.

(2) Preparation of the membrane.
(3) Assembly of the manifold.
(4) Application of the sample.
(5) Blotting.
(6) Processing the membrane.

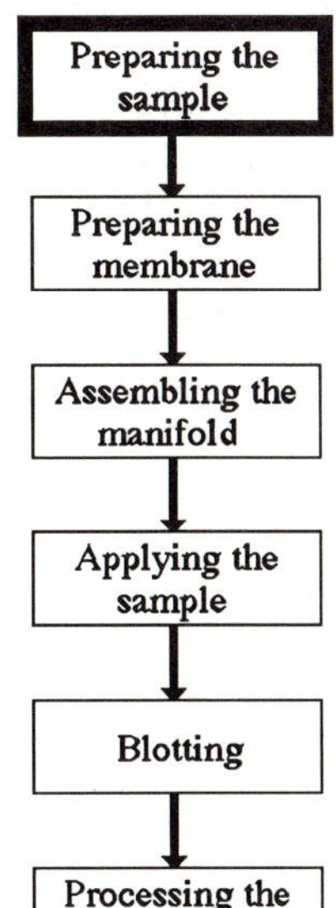

2.1 Preparing the sample

2.1.1 How much DNA or RNA should you load?

The nucleic acid binding capacity of a nylon membrane is in the order of 500 µg/cm^2, whilst that of a nitrocellulose filter is in the order of 100 µg/cm^2. If you assume the average dot/spot to be 0.1 cm^2 in area, the maximum amount of DNA or RNA you should load in a well is 50 µg, for a nylon membrane, and 10 µg, for a nitrocellulose filter. As discussed below, some applications may require you to load more than 10 µg of nucleic acid per dot/slot. This would be yet another reason for using a nylon membrane rather than a nitrocellulose filter for dot/slot blotting (many other reasons are given in Chapter 7, section 1). For many applications, however, you can load less. How much you load depends on the aim of your experiment. For example:

◇ By 'molar excess' we mean that the number of DNA molecules in the dot/slot must exceed the number of corresponding mRNA molecules in the probe.

- If you are using radiolabelled mRNA (or radiolabelled nuclear run-on transcripts) to probe dots/slots of cDNA clones, *the DNA in a dot/slot* must be in molar excess of *the corresponding RNA in the probe*, all of which will then be able to hybridize. This is essential if you want to measure differences in the extent of hybridization to different dots/slots and use the data to estimate differences in the abundance of the corresponding mRNAs. In addition, the greater the amount of nucleic acid on the membrane, the faster will be the hybridization reaction (see volume on hybridization in this series). You should therefore load quite a lot of DNA in each well. It is usual to load 10–40 µg. Although all that really matters is that the DNA in each well is in molar excess, you ought to load equal numbers of molecules in each well. To do this, you must take account of differences in the sizes of the inserts in different cDNA clones.

◇ If you load 40 µg of a cDNA clone that has 3 kb of vector and 1 kb of insert, there will be only 10 µg of insert DNA in the dot/slot.

◇ If you have clone A (3 kb vector + 1 kb insert) and clone B (3 kb vector + 2 kb insert), you can load equal numbers of molecules of the two clones by loading 1.25 µg of clone A for every 1 µg of clone B.

- If you are using a radiolabelled cDNA clone to probe dots/slots of total RNA, and wish to obtain quantitative data, you must load a standard amount of RNA in each well. This requires careful thought, as discussed in Chapter 4, section 3.1, in the context of northern blotting. In this kind of experiment, the radiolabelled cDNA probe has to be in excess, so that there is sufficient probe to bind to all of the corresponding RNA species on the filter. You can therefore load as little RNA in the wells as is necessary to obtain a detectable hybridization signal. For reasons given in Chapter 4, section 3.2.2, it may sometimes be necessary to load a series of dilutions of each sample.

◇ You may have to determine the best amount of RNA to load by trial and error.

2.1.2 DNA samples

In order to hybridize with a probe, the DNA must be denatured. In a Southern blotting experiment, this is usually done by treating the agarose gel with alkali after electrophoresis. In dot/slot blotting, the DNA is usually denatured before it is applied to the membrane. There are two ways in which you can do this:

◇ Refer to the instructions provided by the manufacturer of the membrane you are using.

- denature in alkali—add a 1/10 volume of 2 M NaOH to the sample, leave it at room temperature for 5–15 minutes, neutralize the solution by adding 1.5 volumes of 1.5 M ammonium acetate (pH 4.5), and place it on ice,
- denature by heating—add a 1/10 volume of 20 × SSC to the sample, heat it at 95–100 °C for 5–15 minutes, and place it on ice.

◇ You can use up to 5 M NaOH (final concentration 0.5 M), but should then increase the concentration of ammonium acetate added to neutralize the sample.

◇ Placing the sample on ice slows the rate of renaturation in the time between denaturing the sample and applying it to the membrane.

Both methods work well, but you should note that supercoiled plasmid DNA presents particular problems. First, it may need special treatment to be denatured effectively. For example, you might need to boil the sample for 10 minutes and then incubate it in 0.5 M NaOH for 20 minutes, before neutralizing. In addition, denatured supercoiled plasmid DNA renatures rapidly once the solution is cooled and neutralized, because the separated strands remain physically linked. You must therefore apply the sample to the membrane very soon after cooling/neutralization. It is probably best to linearize plasmid DNA before denaturation, by cutting it with a restriction enzyme, so that supercoiling is no longer a problem.

◇ When supercoiled plasmid DNA is denatured, the separated strands remain physically linked.

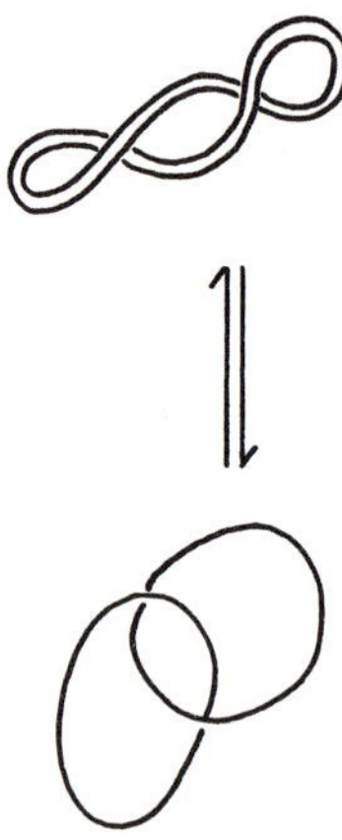

◇ Since the rate-limiting step in renaturation is the collision between complementary strands, supercoiled DNA renatures very rapidly.

If you are using single-stranded DNA, such as M13 phage DNA, then you obviously do not need to denature to separate strands. However, the DNA may contain regions of complementary base sequence, enabling it to form secondary structure by intramolecular base pairing. To eliminate this, single-stranded DNA samples should be treated with alkali or heat, just as for double-stranded DNA.

2.1.3 RNA samples

As discussed previously (Chapter 4, section 1), RNA can also form secondary structure by intramolecular base pairing, and aggregates by intermolecular base pairing. In northern blotting, RNA is denatured before and during electrophoresis, both to ensure that the distance it migrates in the gel is proportional to its length, and to promote efficient hybridization with the probe. For the latter reason, RNA must also be denatured before dot/slot blotting. There are two methods for doing this:

◇ The theory behind this is discussed in Chapter 4, sections 4.2 and 4.3.

- incubate the RNA in 50 per cent (v/v) de-ionized formamide, 6 per cent (v/v) formaldehyde, 1 × SSC at 65 °C for 15 minutes, and place the sample on ice,
- incubate the RNA in 50 per cent (v/v) DMSO, 1 M de-ionized glyoxal, 12.5 mM sodium phosphate, 1 × SSC at 50 °C for 15 minutes, and place the sample on ice.

◇ DMSO attacks nitrocellulose filters. If you must use these, omit the DMSO and increase the incubation time to 1 hour.

There is very little to choose between these methods in terms of their efficacy and ease of use.

2.2 Preparing the membrane

◇ The procedure is essentially the same if you are using a nitrocellulose filter.

◇ Use a sharp scalpel blade, wear gloves, and keep the membrane between the protective sheets while cutting.

◇ Some protocols suggest you soak the wet nylon membrane in SSC. This is not necessary. However, if you are using a nitrocellulose filter it *must* be soaked in 20 × SSC after wetting, since DNA and RNA only bind to nitrocellulose in solutions of high ionic strength.

For the reasons discussed in Chapter 7, section 1, we recommend that you use a nylon membrane, rather than a nitrocellulose filter, for dot/slot blotting.

Cut the membrane to fit the manifold you are using. Nylon membranes *must* be wetted before dot/slot blotting, otherwise capillary action will draw fluid laterally through the membrane during blotting and result in spreading of the sample. Wet the membrane by floating it on de-ionized water (see Chapter 7, section 1.4).

2.3 Assembling the manifold

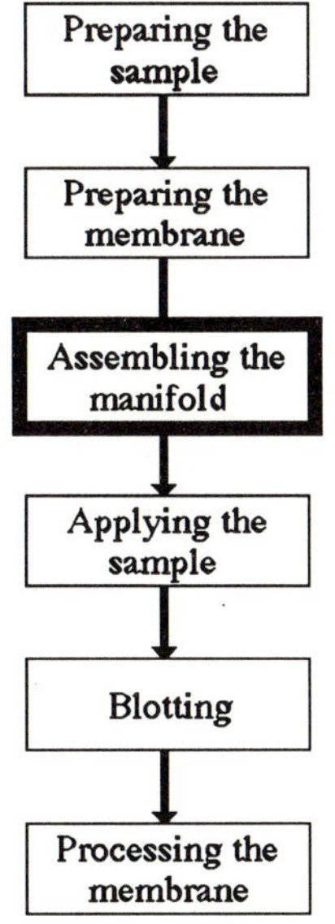

◇ If the manufacturer's recommend that a particular side of the membrane be used to bind DNA and RNA, this face should be next to the block.

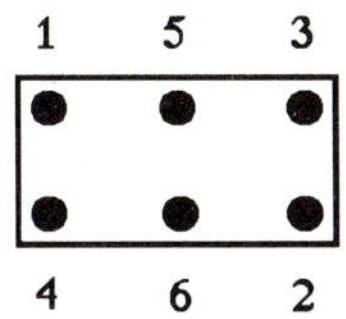

◇ Tighten the nuts shown in the sequence 1-2-3-4-5-6.

If you have not yet purchased a dot/slot blotting manifold, take some time to find one that is easy to assemble and use. One particularly poorly designed model requires the strength of ten and the grappling ability of an octopus, to assemble correctly, but most are within the capabilities of an average person working alone. All manifolds come with instructions as to how they should be used, but these often get lost. A typical manifold would be assembled as follows (*Figure 5.1*):

(1) Lay the upper block on the bench, with its upper face downwards.
(2) Place the wet membrane on top of the block. Smooth away any trapped air bubbles.
(3) Place a sheet of Whatman 3 MM filter paper (cut to size and pre-soaked in the same liquid as the membrane) on top of the membrane. Smooth away any trapped air bubbles.
(4) Place the filter pad, if one is supplied with your blotting manifold, on top of the filter paper.
(5) Place the middle block on top of the filter pad.
(6) Place the lower block on top of the middle block.
(7) Put the screws and nuts in place and clamp the whole thing together, turning it the right way up. To minimize the possibility of leaks, the nuts should all be tightened to the same extent. This is best done by tightening each nut a little at a time, in sequence, until they are all finger tight.
(8) Attach the manifold to the vacuum pump. You can use an electric pump or a tap device.
(9) Fill all the wells with de-ionized water (or whatever you pre-soaked the membrane in), apply gentle suction to the manifold until the liquid in the wells has gone, and then turn off the suction.

Watch out for the following problems:

- Check that all the liquid has gone from every well. You often find wells that do not drain properly and it is best to identify these so that you can either remedy the problem or avoid using them.

◇ This is particularly important if you have not used the manifold before, or for a while, since damage to the manifold can cause samples to spread.

◇ With most manifolds, you can stain the membrane, examine it without dismantling the manifold, and then perform your dot/slot blot on the same membrane.

- If your samples spread out from the dot/slot in any of the wells, you will end up with hybridization signals like those in *Figure 5.4B*. To check for this, add electrophoresis gel loading buffer (Chapter 2, section 2) to the wells and suck it through so that the dye in the buffer stains the membrane. Examination of the membrane will show whether the stain has the correct dot or slot shape, or whether it has spread. In our experience, the dye does not affect binding of DNA or RNA to the membrane, or hybridization. Since the membrane concentrates the dye by removing it from solution, you should use a 1/100 dilution of your concentrated stock of electrophoresis gel loading buffer (Chapter 2, section 2). If the samples do spread from any of the wells, check that the manifold is assembled properly. If you cannot prevent spreading, avoid using the wells that are giving you problems.

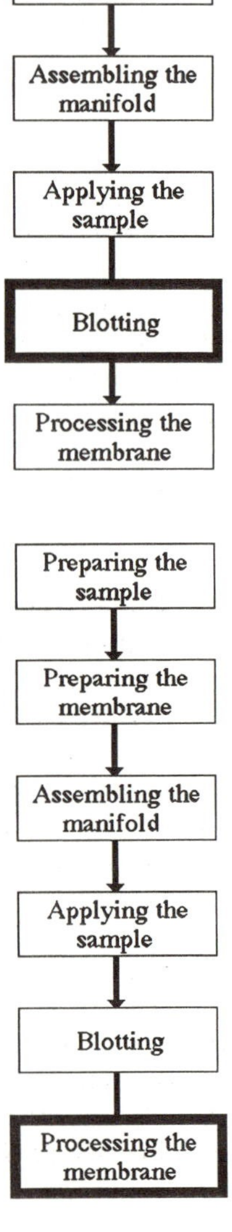

2.4 Applying the sample

Load the samples into the slots. Keep a record of what went where.

2.5 Blotting

Opinion is divided as to how dot/slot blotting should be performed. These are the options:

- apply a gentle vacuum as soon as all the wells are loaded,
- leave the samples to stand in the wells for anything from 10 minutes to 2 hours and then apply a gentle vacuum.

Both methods seem to work well. The former is obviously quicker. In either case, you should apply the vacuum until all the liquid has been sucked from every well. This may take only 30 seconds. Some people chase the sample through by adding 1 ml of de-ionized water (or whatever you pre-soaked the membrane in) to each well and sucking it through. Some people do this twice, but it is probably not necessary to do so.

Finally, loosen the nuts that hold the manifold together, take the manifold to pieces and carefully remove the membrane. Some people rinse the membrane in 2 × SSC at this stage.

2.6 Processing the membrane

The method for processing the membrane depends on the nature of your samples:

- DNA samples—fix the DNA to the membrane as described in Chapter 3, section 1.5.
- RNA samples denatured in formaldehyde—bake the membrane at 80 °C for 2 hours to reverse the formaldehyde reaction. This

should also fix the RNA covalently to the membrane. If the manufacturer recommends UV cross-linking, then you should do this as described in Chapter 3, section 1.5.

◇ Do ***not*** treat nitrocellulose filters with sodium hydroxide. It makes them very brittle and thus difficult to handle.

- RNA samples denatured in glyoxal—reverse the glyoxal reaction, ***either*** by heating the membrane for 5 minutes at 100 °C in 20 mM Tris-HCl, pH 8.0, ***or*** by soaking the membrane in 50 mM NaOH for 15 seconds and then in 1 × SSC, 0.2 M Tris-HCl, pH 7.5, for 30 seconds. Finally, fix the RNA to the membrane by air drying or UV cross-linking, as described in Chapter 3, section 1.5.

3. Quantitation of dot/slot blots and interpretation of results

You will often want to compare the amount of probe that has hybridized to different samples. This can be done by:

- scintillation counting of excised dots/slots,
- scanning densitometry,
- phosphorimaging.

The advantages and disadvantages of these methods are discussed in Chapter 4, section 3.2. How you handle the data generated by these methods depends on the aim of your experiment:

◇ The DNA in the dot/slot must be in molar excess, as described above.

- If you have probed dots/slots of cDNA clones with mRNA that has been radiolabelled at its 5′ end, handling the data is straightforward—if one dot/slot has bound twice as much radioactivity as another, then there are twice as many mRNA molecules bound to it.
- If you have probed dots/slots of cDNA clones with radiolabelled nuclear run-on transcripts, then matters are more complicated. This is because such probes are not labelled just at one end, but are continuously labelled throughout their length. The amount of radioactivity in a given RNA species in the probe is therefore proportional to its length. Suppose you are comparing hybridization to dots/slots of clone A (with a 2 kb insert) and clone B (with a 1 kb insert). If the probe contains equal numbers of radiolabelled transcripts corresponding to the two clones, equal numbers of ***molecules*** will bind to the two clones, but there will be twice as much ***radioactivity*** bound to clone A. You must therefore correct your data to account for differences in the lengths of the target cDNA sequences in each dot/slot.
- If you have probed dot/slots of RNA with a radiolabelled cDNA probe and want to relate differences in the amount of probe hybridized to each dot/slot to differences in the abundance of a the corresponding mRNA, you must take account of the issues that were discussed in Chapter 4, section 3.2, in relation to northern blotting (*Figure 5.5*). You must also include the controls described there.

◇ The same is true for cDNA probes synthesized from mixed mRNA populations.

◇ In fact, this interpretation rests on a number of assumptions about the nature of the cDNA clones and the behaviour of nuclear run-on transcripts in such experiments. We will not discuss these details here.

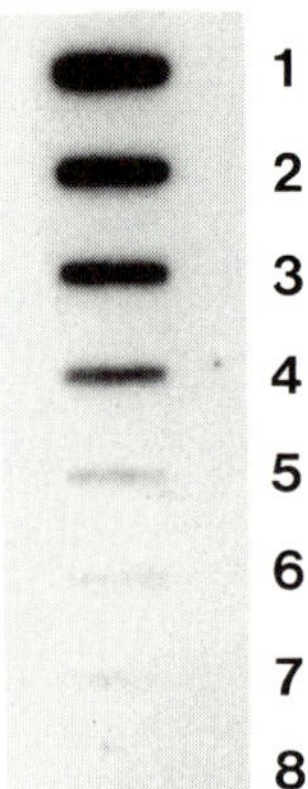

Fig 5.5

A problem with scanning densitometry (see Chapter 4, section 3.2.2) is that the response of X-ray film may be non-linear. A series of 10-fold dilutions of a DNA sample were slot blotted on to nylon, hybridized and autoradiographed. Subsequent scintillation counting of excised 'slots' showed that there was 10 times more radioactivity bound to each sample than to the sample below it. This is only reflected accurately in the intensity of some of the autoradiographic signals. For example, signal 4 is 10 times as intense as signal 5, but signal 1 is **not** 10 times as intense as signal 2. Only signals 4, 5, and 6 lie within the film's linear range

4. Limitations of dot/slot blotting

◇ Dots/slots may hide a multitude of sins.

◇ You cannot tell if your dot/slot blotted RNA is degraded, without first checking it by agarose gel electrophoresis and, preferably, by northern blotting.

A number of the difficulties in using this technique have been discussed above. However, probably the greatest limitation is that you cannot be sure what your probe is hybridizing to in a dot/slot. For example, if you probe a northern blot with a radiolabelled cDNA probe, you can see whether the probe hybridizes specifically to the mRNA of interest, or whether there is significant non-specific hybridization. For example, probes often hybridize non-specifically to ribosomal RNA on northern blots (*Figure 5.6*). If the same thing happens on your dot/slot blot, the value of your data will be seriously compromised. It is therefore crucially important that you characterize your probe fully by testing its specificity on a northern blot before you use it to probe dot/slot blots. Similar considerations apply to the other types of dot/slot blotting experiments described in this chapter.

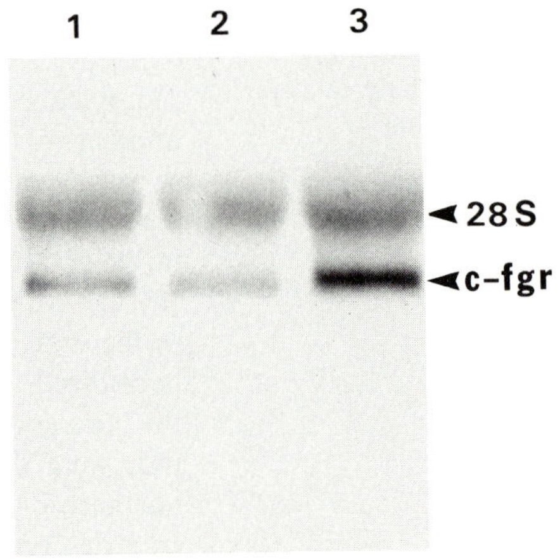

Fig 5.6

Northern analysis of c-*fgr* mRNA in total RNA from 3 B-lympoma cell lines. The probe hybridized to c-*fgr* mRNA (3 kb) and cross-hybridized to 28S rRNA. There are clear differences in c-*fgr* mRNA levels in the 3 cell lines, but it would be impossible to quantitate these on a dot/slot blot because of the cross-hybridization to 28S rRNA

5. Further reading

Costanzi, C. and Gillespie, D. (1987). Fast blots: immobilization of DNA and RNA from cells. *Methods in Enzymology*, **152**, 582–7.

Perbal, B. (1988). *A practical guide to molecular cloning*, (2nd edn), pp. 438–9, Wiley, New York.

◇ RNA dot/slot blots only.

Sambrook, J., Fritsch, E.F. and Maniatis, T. (1989). *Molecular cloning: a laboratory manual*, (2nd edn), Vol. 1, pp. 7.53–7.57, Cold Spring Harbor Laboratory Press.

6. References

Kafatos, F.C., Jones, C.W., and Efstratiadis, A. (1979). Determination of nucleic acid sequence homologies and relative concentrations by a dot blot hybridization procedure. *Nucleic Acids Research*, **7**, 1541–52.

Lillycrop, K.A., Dent, C.L., and Latchman, D.S. (1992). Regulation of gene expression in neuronal cell lines. In *Neuronal cell lines: A practical approach*, (ed. J.N. Wood), pp. 181–215, IRL Press at Oxford University Press.

Marzluff, W.F. and Huang, R.C.C. (1985). In *Transcription and translation: A practical approach*, (ed. B.D. Hames and S.J. Higgins), pp. 89–129, IRL Press at Oxford University Press.

6 Plaque and colony screening

You may want to search through collections of different cloned DNA molecules in order to find a particular molecule of interest. Your cloned DNA may be in the form of recombinant bacteriophage λ particles or of bacteria that contain recombinant plasmids or cosmids. Your clones may be subcloned fragments of a larger piece of DNA under study or may represent a complete genomic or cDNA library. In this chapter we will discuss how you can immobilize your collection of cloned DNA fragments, ready for screening by hybridization with a labelled nucleic acid probe.

As well as screening for the cloned DNA fragment of interest by hybridization, it is possible to screen for the presence of the protein encoded by the clone. To do this, you could use an antibody, ligand or protein-binding oligonucleotide, specific for the protein you are looking for. Immobilization of protein from clones, for screening by these methods, will not be discussed in this book. For further information about these techniques, you might consult Cowell and Hurst (1993), Helfman and Hughes (1987), and Mierendorf *et al.* (1987).

◇ There are many other considerations that will influence the decision whether to use a bacteriophage λ, plasmid or cosmid library.

As you will soon discover, it is technically much easier to screen collections of bacteriophage λ plaques than collections of bacterial colonies containing plasmids. This is one of the advantages of using cDNA and genomic libraries constructed in bacteriophage λ vectors rather than in plasmid or cosmid vectors.

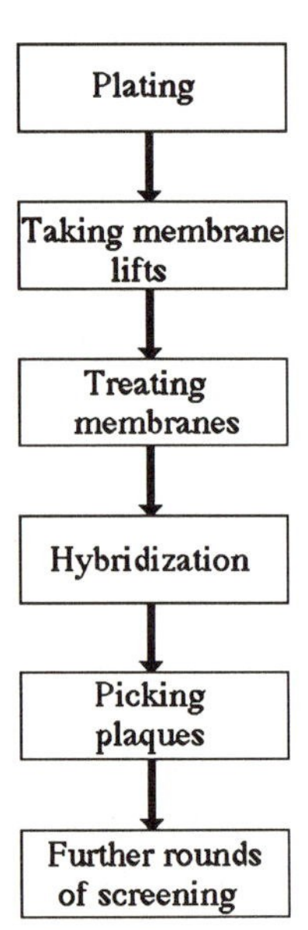

1. Screening bacteriophage λ plaques by the Benton and Davis method

The most commonly used method for screening collections of bacteriophage λ plaques was developed by Benton and Davis (1977). In this procedure, the collection of recombinant bacteriophage is plated out on a lawn of *E. coli* in agarose. Each bacteriophage λ particle will infect a single bacterium, replicate inside, and subsequently lyse it, releasing many identical bacteriophage. In semi-solid agarose, diffusion of these will be limited and they will only infect bacteria immediately surrounding the lysed bacterium. These bacteria will in turn be lysed and bacteriophage released from them will infect bacteria surrounding them. After successive rounds of infection, regions of lysed bacteria become visible as a cleared areas, or plaques, in the opaque

A

B

C

D

Fig 6.1

Formation of plaques by bacteriophage λ in a lawn of *E.coli* cells (white boxes). In **A**, a cell (shaded) is infected by a λ particle. In **B**, this cell lyses (black) and releases infectious λ particles that infect surrounding cells. In successive rounds of infection and lysis, (**C** and **D**), the lysed area grows in size, becoming visible as a cleared area, or plaque, in the bacterial lawn

◇ A single plaque contains millions of identical bacteriophage λ particles.

◇ This is called a 'membrane lift'.

◇ Non-radioactive probes can also be used very successfully.

lawn of bacteria (*Figure 6.1*). Once the plaques have developed sufficiently, usually after overnight incubation, you place a nylon membrane on top of the agarose so that bacteriophage λ particles from each plaque are transferred from the plate to the membrane. When you remove the membrane, it carries a replica of the pattern of plaques on the plate. You then treat the membrane to break open the bacteriophage λ particles, to denature the DNA they contain and to fix the DNA to the membrane. You can then detect specific DNA sequences on the membrane by hybridization with a radiolabelled probe, followed by autoradiography. Hybridizing plaques are identified by aligning the autoradiogram with the original agarose plate and can be picked for further analysis.

1.1 Plating

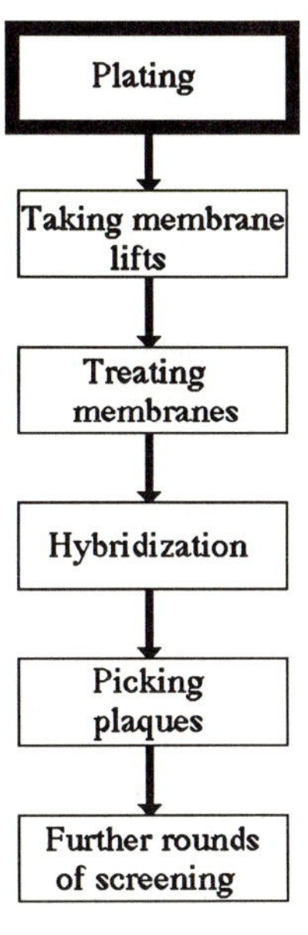

Before plating out your bacteriophage λ, you must prepare:

(1) Petri dishes containing set base agar,
(2) molten top agarose,
(3) plating cells,
(4) a dilution of your bacteriophage λ stock.

The following sections will tell you how to prepare and use these items.

1.1.1 Petri dishes

Use plastic Petri dishes (*Figure 6.2*). The size of dish you use is determined by the number of plaques you wish to screen. If you have only a small collection of bacteriophage λ clones, plate them at low density so that the positions of individual plaques can be easily distinguished. For example, you could plate around 100 bacteriophage λ particles on a standard 90 mm circular Petri dish. If you are screening a large collection of clones, such as a cDNA or genomic

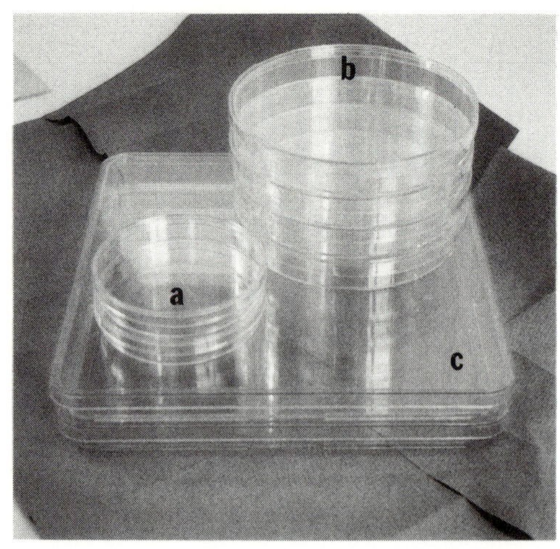

Fig 6.2

Plastic Petri dishes: a, 90 mm circular; b, 150 mm circular; c, 20 × 20 cm square

◇ If you want to screen 200 000 plaques, one 20 × 20 cm dish is easier to handle than 20 90 mm dishes.

library, plate the bacteriophage particles at high density, otherwise you will find yourself handling an unmanageable number of dishes. In high density screening, the dish should be completely covered with plaques, so that hardly any unlysed bacteria can be seen between them. However, there is a limit to the number of plaques you should plate on a single dish:

- If you plate at too high a density, the plaques will be too small because they can only grow until they touch each other. As a result, you will get weak hybridization signals.
- When you pick a plaque from a high density plate it is bound to be contaminated with neighbouring plaques and you will have to perform further rounds of screening to purify the plaque you want (see section 1.8). If you plate at too high a density, the level of contamination by neighbouring plaques is so great that it becomes difficult to find a hybridizing plaque in subsequent rounds of screening.

Since there is variation in the size of plaques produced by different strains of bacteriophage λ, the number of plaques that will just cover the dish will not always be the same. However, *Table 6.1* gives a rough guide to the numbers you should aim at.

◇ Plastic dishes cannot be autoclaved. They melt.

◇ To avoid later confusion, remove all traces of marker pen from the dishes with alcohol.

The dishes must be sterile. They are best used as they come from the manufacturer and then discarded. However, since 20 × 20 cm dishes are expensive we re-use them. Re-use carries a slightly increased risk of contamination, so take steps to minimize this. Wash used dishes carefully to remove all visible traces of agar and bacteria. Then rinse them in distilled water, sterilize by immersion in industrial methylated spirits, and dry in a hood, preferably under UV radiation.

1.1.2 Base agar

◇ Use good quality microbiological agar.

To prepare Petri dishes containing solid base agar, first add agar to L broth (1% (w/v) tryptone, 0.5% (w/v) yeast extract, 0.5% (w/v) sodium chloride) to give a concentration of 1.1 per cent (w/v). The L

◇ It is advisable to add maltose but not, in fact, essential.

broth should contain 10 mM magnesium sulphate, since magnesium ions are essential for assembly of bacteriophage λ particles. Most protocols also suggest that you include 0.2 per cent (w/v) maltose in the broth. This is to induce the *Escherichia coli* maltose operon and so increase levels of the *lam*B gene product, which is the receptor used by bacteriophage λ particles to bind to and enter *E. coli* cells. Autoclave the mixture to dissolve the agar and to sterilize the solution, allow it to cool to around 60 °C, and pour it into Petri dishes. If the mixture is too hot when you pour it, it may distort the dishes, but if it cools too much it will start to set and give lumpy plates. Add enough agar to give a reasonably thick layer that will not dry out quickly, but be as economical as possible (*Table 6.1*). We add about:

◇ Protocols frequently over-estimate the amount of base agar that should be added to a dish.

◇ It is possible to get away with less.

- 15 ml to a 90 mm dish,
- 50 ml to a 150 mm dish,
- 250 ml to a 20 × 20 cm dish.

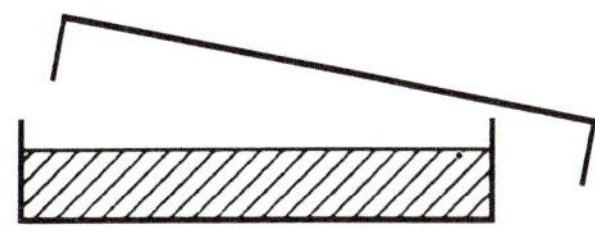

Leave poured dishes to set on a level surface with their lids cocked. It is best to do this in a relatively quiet corner of the lab, to minimize contamination by airborne microbes. Once the agar has set completely, replace the lids, wrap the dishes in cling film and store at 4 °C. Store the plates inverted so that droplets of condensation form on the inside of the lid rather than on the agar, which would encourage contamination. When you use the plates, carefully wipe away any condensation in the lids.

◇ Set agar is slightly more opaque than molten agar and should not wobble when you tap the side of the dish.

Condensation can also cause problems when the dishes are subsequently being incubated at 37 °C to allow bacteriophage growth. Small droplets of water can condense on the surface of the top agarose and cause the developing plaques to streak and run into each other. For this reason, it is actually better **not** to use freshly poured plates for plating bacteriophage. Storage for a couple of days allows some of the moisture to evaporate from the top of the agar and helps to avoid these problems. Whether you use stored or freshly prepared plates, place them in an incubator at 37 °C for at least 2 hours immediately before use to promote further drying. Lay the plates out singly, rather than in piles, and invert them, with each plate cocked at an angle against the side of its lid. You can speed up the drying

◇ Drying a plate also prevents the top agarose layer peeling off when a membrane is lifted from its surface.

Table 6.1 Plating out bacteriophage λ particles

Type of Petri dish	90 mm circular	150 mm circular	20 × 20 cm square
Approximate maximum number of plaques	10 000	50 000	200 000
Volume of base agar (ml)	15	50	250
Volume of top agarose (ml)	5	10	50
Volume of plating cells (ml)	0.1	0.3	1
Maximum volume of bacteriophage λ suspension (ml)	0.1	0.3	1

process by increasing the temperature of the incubator to 45 °C but take care not to leave the plates drying for too long, or else the base agar will dry out until it is wafer-thin. If you have had to prepare fresh plates immediately before use and have not had time to dry them properly, place a piece of dry filter paper inside the lid of each plate during incubation. This will reduce the humidity inside the dish and will reduce streaking. The top agarose layer might still peel during membrane lifting, however, and so it is best to plan ahead and use properly dried plates.

1.1.3 Top agarose

The bacteriophage λ particles will be mixed with *E. coli* and spread on top of the base agar in a layer of agarose. We use agarose for the top layer, because agar top layers are much more likely to peel off when taking membrane lifts. Dissolve the agarose by autoclaving in L broth containing 10 mM magnesium sulphate and 0.2 per cent (w/v) maltose. The final concentration of agarose should be 0.75 per cent (w/v). Prepare molten top agarose some time before you want to use it because it must be allowed to cool before bacteriophage and *E. coli* are mixed with it. If the top agarose is too hot when this is done, the bacteriophage and bacteria will be destroyed. However, if it is too cool, you will find it hard to spread a smooth layer of top agarose on the base agar without it setting and giving lumps. Most protocols recommend that you cool the top agarose to 45 °C and maintain it at this temperature until you are ready to use it. If you are experienced at plating, this is fine. If you are inexperienced, and especially if you are plating on to a 20 × 20 cm dish, you will find it easier to use top agarose that is hotter.

◇ Use good quality microbiological-grade agarose.

◇ We use top agarose at 60 °C without noticing detrimental effects on bacteriophage or bacteria.

You will obviously need enough top agarose for all the plates you are using (*Table 6.1*). We use:

- 5 ml for a 90 mm dish,
- 10 ml for a 150 mm dish,
- 50 ml for a 20 × 20 cm dish.

You can use smaller volumes, but if you do it will be more difficult to get a smooth top layer before the agarose sets.

◇ It pays to prepare more top agarose than you need. It is stressful to pour plates from a molten stock that is rapidly depleting and beginning to set in the bottle.

1.1.4 Plating cells

Make sure that you use the correct strain of *E. coli* for growing the strain of bacteriophage λ that you have. This can be done by reference to the supplier's literature (if the strain has been obtained commercially), from the supplier (if in another laboratory), or from an appropriate laboratory manual, such as Sambrook *et al.* (1989).

Most protocols suggest you prepare stocks of plating cells by growing the *E. coli* to mid-log phase, pelleting the bacteria and resuspending them in 10 mM magnesium chloride at an OD_{600} of 2. We do not do this because we find that fresh overnight cultures of *E. coli* work perfectly well and we are reluctant to risk contaminating the cells by manipulating them more than necessary. In fact,

◇ Remember to grow the *E. coli* in L broth containing 10 mM magnesium sulphate and 0.2 per cent (w/v) maltose.

◇ This corresponds to approximately 1.6×10^9 cells per ml.

although we probably ought not to admit it, we find that overnight cultures of *E. coli* stored at 4 °C for as long as 2 weeks work perfectly well. This approach may give poorer plaque morphology, which may cause problems when screening some of the older, weaker strains of bacteriophage λ, such as those of the Charon series. However, we get good results with more vigorous strains such as λgt10 and λgt11 (Huynh *et al.* 1984), the EMBL series (Frischauf *et al.* 1983), and the λZAP series (Short and Sorge 1992).

The quantity of plating cells needed depends on the size of the dish (*Table 6.1*). We use:

- 0.1 ml for a 90 mm dish,
- 0.3 ml for a 150 mm dish,
- 1 ml for a 20 × 20 cm dish.

1.1.5 Bacteriophage λ

We assume that you have a stock suspension of bacteriophage λ particles of known titre (or concentration). If you do not have a reliable estimate of the titre of your stock you will have to determine this yourself. Methods for doing this are described in section 2. Immediately before use, prepare an appropriately diluted suspension of bacteriophage λ particles in SM buffer. In deciding the concentration of bacteriophage λ particles in this diluted suspension you must take account of the size of the dishes to be used and the number of plaques that you want on each. For each plate, the volume of bacteriophage suspension should not exceed the volume of plating cells (*Table 6.1*). For example, if you want 10 000 plaques on a 90 mm dish your diluted suspension should contain 10 000 bacteriophage particles in not more than 0.1 ml.

◇ SM buffer contains 100 mM sodium chloride, 10 mM magnesium sulphate, 50 mM Tris-HCl, pH 7.5, 0.01 per cent (w/v) gelatin. Gelatin stabilizes bacteriophage λ particles.

◇ It does not matter if the suspension is more concentrated than this.

◇ If you plate from a precious stock of bacteriophage, such as a genomic or cDNA library, do not prepare any more of the diluted suspension than you need. Dilute suspensions do not store well.

1.1.6 Plating

When you have prepared dry base agar dishes, molten top agarose at the correct temperature, plating cells and a bacteriophage λ suspension of the correct titre, you are ready to plate out the suspension.

Mixing the ingredients

Using sterile micropipette tips, add the appropriate volume of plating cells and the appropriate volume of bacteriophage λ suspension to a sterile plastic tube of appropriate size. We use:

- 5 ml sterile plastic bijoux for plating on to 90 mm dishes,
- 25 ml sterile plastic universal containers for 150 mm dishes,
- 50 ml sterile plastic Falcon tubes for 20 × 20 cm dishes.

Cap the tubes and place them at 37 °C for 20 minutes, to allow the bacteriophage λ to attach to the surface of the *E. coli*. Towards the end of this period, remove the base agar dishes from the incubator where they are drying. Number them on their bases using a water-fast marker pen. Place the dishes on a flat surface, singly, rather than in piles, with their lids on and uppermost.

◇ Do not label the lids, as these can subsequently be switched between dishes by mistake.

Remove the bacteriophage/*E. coli* mixtures from the 37 °C incubator and retrieve the molten top agarose from its incubator. If the top agarose is at 45 °C and you have a lot of plates to pour, you will probably need to keep it beside you in a 45 °C water bath, to prevent it setting during pouring. If you start with top agarose at 60 °C and/or have only a few plates to pour you will probably be able to manage with the the bottle of top agarose standing on the bench top beside you. Taking one tube of bacteriophage/*E. coli* at a time, add the appropriate volume of top agarose. It is best to do this by pouring directly from the bottle of top agarose into the tube and estimating the amount that you have added by eye. You can use sterile glass or plastic pipettes to aliquot the top agarose, but this gives extra time for it to set, extra opportunity for contamination, and extra washing up (or cost, if you use disposable pipettes). If you do pour the top agarose, however, make sure that you do not allow the top of the bottle to become contaminated with the contents of the tube into which you are pouring.

◇ It is not necessary to add exactly the stipulated amount.

Pouring the plates

Once the top agarose is added, swiftly but smoothly cap the tube, mix the contents by gently inverting it, uncap the tube and pour the contents on to the surface of the base agar in one of the dishes. Do not be timid about pouring the top agarose on to the dish. If you pour boldly, you will find that it spreads more widely across the surface of the base agar. Encourage the top agarose to spread smoothly over the whole of the base agar surface by gently tilting the dish from side to side. Do all this as quickly as possible, but avoid generating air bubbles in the top agarose. In particular, *do not* shake the last dregs of top agarose on to the dish. These are likely to contain air bubbles but will not contain a significant proportion of your bacteriophage. If there are air bubbles in the molten top agarose after it has been added to the dish, pop them with a sterile micropipette tip or sterile toothpick. When you are happy with the poured dish, replace its lid, slightly cocked, and pour the next dish.

◇ Having warm base agar plates also helps to spread the top agarose.

◇ Once the top agarose begins to set, it is best to leave well alone and live with any bubbles that remain.

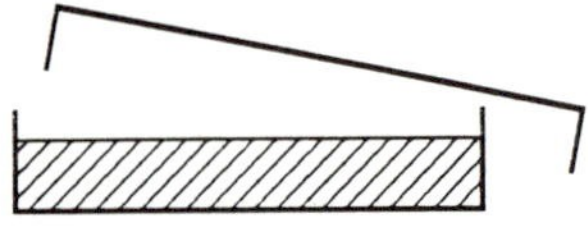

The above procedure can be difficult at first, but with practice you will learn the knack of quickly pouring large numbers of smooth, bubble-free plates. Warm base agar dishes, good volumes of hot top agarose and a minimum of fussy pipetting will help you to achieve this while you gain experience.

◇ If you have never poured plates before, practice using SM buffer in place of bacteriophage.

Incubating the plates

When you have poured all the plates, leave them until the top agarose has set thoroughly. The larger the dish and the hotter the day, the longer this will take. If you are using 20 × 20 cm dishes, leave them for 30 minutes to be quite sure they have set. When the plates are ready, replace the lids fully, invert the plates carefully and incubate them, inverted, at 37 °C overnight. If your incubator is large enough, lay the plates out singly rather than in piles. We feel that this further reduces the risk of condensation.

◇ Top agarose that has not set completely will slide into the lid when you invert the plate. It does so rather gracefully, but this is little compensation for the pain it will cause you.

What should you see the morning after?

After overnight incubation, inspect the plates. A typical plate is shown in *Figure 6.3*. If the number of plaques is of the correct order

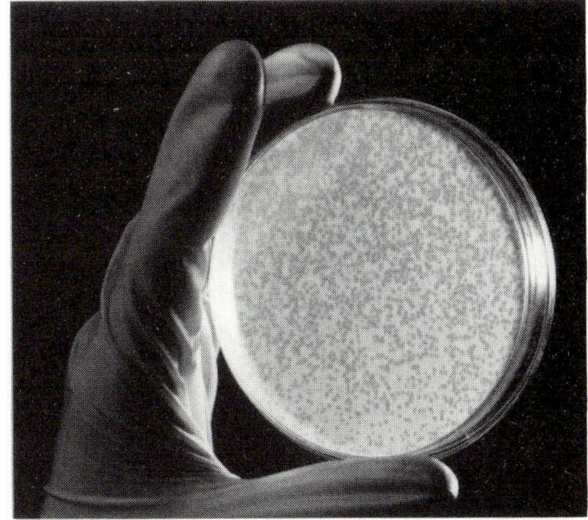

Fig 6.3

Bacteriophage λ plaques in a lawn of *E.coli* in a 90 mm plastic Petri dish. (Photograph courtesy of Amersham International plc)

◇ Chilling the plates further reduces the risk of peeling or fragmenting of the top agar when taking membrane lifts.

and if the plates are free from contamination, place the plates, inverted, in a refrigerator or cold-room for an hour before you take membrane lifts.

When performing high density screening the plaques should cover the plate so that they touch each other but should still leave tiny areas of unlysed bacteria between them. For the reasons discussed above, plates that are completely confluent are probably not worth proceeding with. If there are far too few plaques on the plate it might be worth proceeding. On one occasion we miscalculated and plated 2 000 plaques of a genomic library on a 20 × 20 cm dish instead of 200 000. Feeling lucky, we screened membranes taken from the plate and found the clone we wanted, although the odds were heavily against. Since the library was plated at such low density we were able to pick a pure plaque with no need for further rounds of screening.

◇ This does not always happen.

Similarly, contaminated plates may be useable. Contamination *in* top agarose or base agar is a real problem and you should discard the plates. On several occasions, however, we have rescued plates that have grown a thick creamy layer of contaminant on the surface of the top agarose. If you gently scrape such a layer away with the edge of a clean microscope slide, taking care not to damage the top agarose, you may find perfectly good plaques in the *E. coli* lawn beneath. We have successfully screened such plates. Of course, it is better not to have contamination and you might believe that it is never worth continuing with a half-ruined experiment. *In extremis*, however, it is worth knowing that plates contaminated in this way are redeemable.

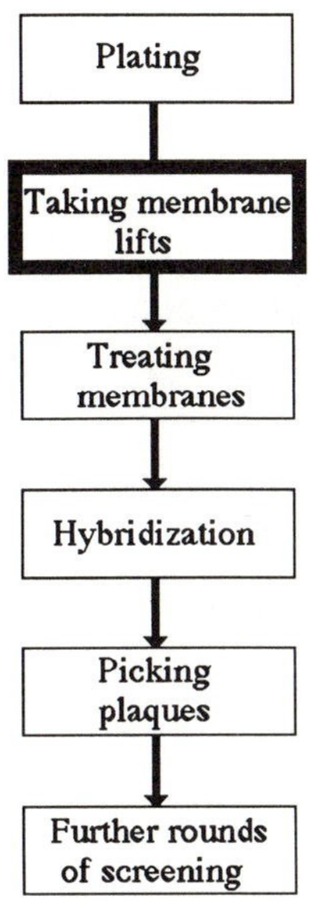

1.2 Taking membrane lifts

1.2.1 Getting the membranes ready

While the plates are chilling, prepare space in which to work. You will need enough bench space to lay all the membranes out beside each other. You should also prepare the membranes, remembering to handle them with gloved hands at all times. As discussed in Chapter 7, section 1, nylon membranes are far superior to nitrocellulose filters

◇ For hints on using nylon membranes and nitrocellulose filters, see Chapter 7, section 1.

◇ Spots of spurious probe binding are a serious problem in plaque/colony screening, because real signals are also spots. They are less trouble in Southern/northern blotting, where real signals are bands.

for taking lifts from bacteriophage λ plates, if you are planning to screen the filters by nucleic acid hybridization. You should use nitrocellulose filters only if you are screening with an antiserum. Which nylon membrane to use is largely a matter of personal choice. Some manufacturers supply circular membranes cut to fit 90 mm and 150 mm circular dishes. Unless you have a gadget for cutting circles from sheets of membrane it is worth paying for these. It is very laborious to cut circles with scissors or a scalpel blade and you risk damaging the surface of the membrane and so giving rise to spurious binding of the probe during hybridization. It is also possible to buy ready-cut 20 × 20 cm membranes, although it is obviously easy to cut squares yourself. Use membranes as they come from the manufacturer. It is unnecessary and inadvisable to wet or sterilize them before use.

◇ Artefactual signals will not, in general, be duplicated.

You will need two membranes for each plate because it is *absolutely essential* to take duplicate membrane replicas of each plate. It is common to obtain a few random spots on an autoradiograph of a hybridized membrane as a result of artefactual binding of the probe during hybridization. Since such artefactual signals may resemble real hybridization signals, the only way to distinguish the real thing is by observing a signal at the same position on two duplicate membranes.

1.2.2 Taking the first membrane lift

When you are ready, remove the plates from the refrigerator or cold-room. Take one plate and proceed as follows.

◇ Wear disposable gloves at all times.

◇ If the manufacturers recommend that a particular side of the membrane be used to bind DNA, place this side down on to the top agarose.

Using a soft lead pencil, label the edge of a membrane with the number of the plate and the letter 'A' to show that this is the first membrane to be taken from the plate. Hold the edges of the membrane so that its centre bends down slightly under its own weight (*Figure 6.4A*). Allow the centre of the membrane to contact the centre of the plate. The membrane will immediately become wet at the point of contact. It is then very easy to lower the rest of the membrane gently on to the surface of the top agarose (*Figure 6.4B*). If you lay the membrane down in this way, from the centre outwards, it will

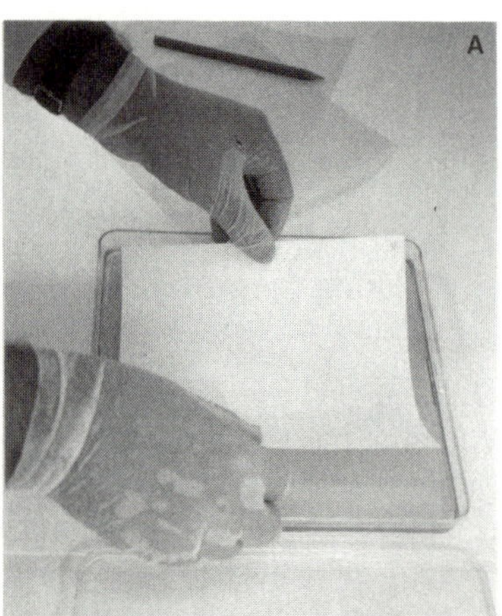

Fig 6.4

Lowering a nylon membrane on to a lawn of *E. coli* containing bacteriophage λ plaques

◇ Do not proceed with a membrane that does not make proper contact over its whole area. Throw it away and use another.

◇ The needle may become blunted, so that it catches on the membrane as you pull it out. Have a stock of new needles close to hand.

◇ Make the pen marks small, to enable accurate alignment later.

◇ The recommended time for transfer to the first membrane is 1–2 minutes.

make contact over its whole area, without creasing or trapping air bubbles. *Under no circumstances* move the membrane sideways, or remove it and replace it, once the initial contact with the plate has been made.

When the membrane is down, orientate it to the plate by making marks at corresponding positions on the plate and membrane (*Figure 6.5*). To do this, take an 18 gauge hypodermic syringe needle and gently stab it vertically downwards through membrane and agar and into the bottom of the dish. Carefully remove the needle and repeat the process in another place so that you have a series of orientation marks, such as those shown in *Figure 6.6*. The disposition of the orientation marks must be asymmetric so that the position of the membrane on the plate can later be determined accurately and unequivocally.

With the membrane still on the plate, replace the lid, invert the plate, and mark the positions of the needle marks on the bottom of the dish using a water-fast marker pen (*Figure 6.7*).

By the time you have done all of this, sufficient bacteriophage from each plaque will have adhered to the membrane. Carefully lift one edge of the membrane with gloved fingers or blunt-ended forceps and peel it off. Place it, with the bacteriophage side uppermost, on a sheet of low grade filter paper (*Figure 6.8*).

Fig 6.5

Making an orientation mark in a Benton-and-Davis membrane with an 18 gauge hypodermic syringe needle

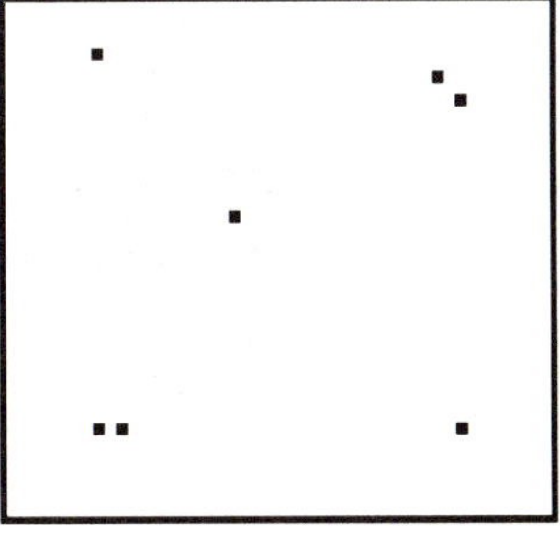

Fig 6.6

A typical asymmetric pattern of orientation marks

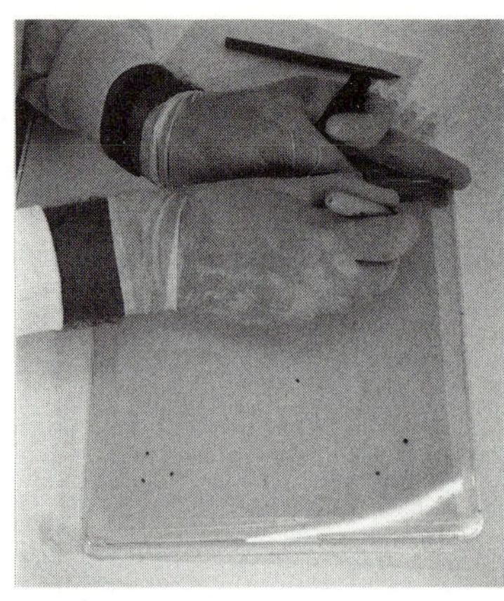

Fig 6.7

Marking the positions of the orientation marks on to the bottom of the Petri dish, using a water-fast marker pen

Fig 6.8

Lifting a membrane from a plate, with blunt forceps. These bacteriophage λ plaques are blue because the bacteriophage carry the *E.coli lacZ* gene. Infected *E.coli* therefore express β-galactosidase, which generates a blue product when X-Gal is in the top agarose. (Photograph courtesy of Amersham International)

1.2.3 Taking the second membrane lift

Take a second membrane, label it with the number of the plate and the letter 'B' and lower it on to the surface of the plate as before. Take an 18 gauge needle and pierce the membrane at the exact positions of the pen marks on the bottom of the dish. ***Absolute accuracy is crucial.*** When you have marked the membrane, replace the lid on the plate and place it to one side. Leave the second membrane on the plate until you have dealt with all the other plates.

◇ You may find it easier to do this if you place the dish on a light box.

◇ The length of time the second membrane should be in contact with the plate is not crucial, as long as it exceeds 2 minutes.

Repeat the above procedure for each of the remaining plates. When all the plates have been dealt with, remove the second membranes from them, placing them on filter paper.

Some people take several membrane lifts from one plate. However, you will find that the hybridization signal intensity drops dramatically after the third membrane and it is rarely worthwhile taking more than two.

Some people mark the positions of the needle holes on the membranes using biro, water-fast marker pen or India ink. Others rely on their ability to find the tiny needle marks at a later stage. This is a matter of personal choice.

1.2.4 Storing the plates

When you have removed all the membranes, wrap the plates in cling film or seal their lids with parafilm, so that they do not dry out, and store them inverted in a refrigerator or cold-room. Taking membrane lifts is bound to introduce contaminants to the plates and you will find that after a few weeks of storage many exotically coloured colonies develop. You should therefore proceed swiftly with hybridization so that you can pick interesting plaques long before contaminants become a problem.

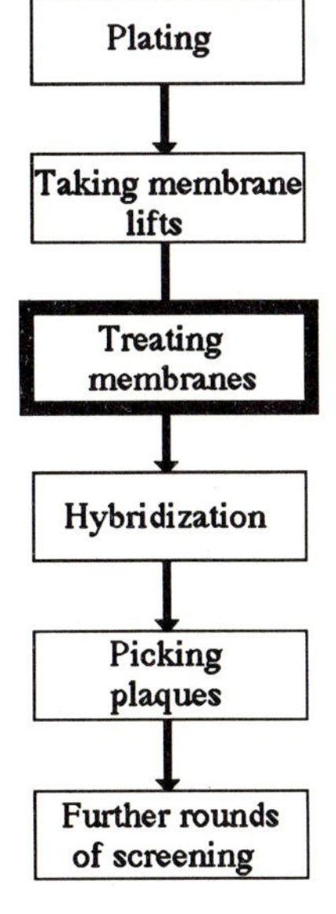

1.3 Treating membranes before hybridization

You must now treat the membranes with denaturation solution, neutralization solution and 2 × SSC, and then fix the DNA to the membrane.

1.3.1 Denaturation

Denaturation solution breaks open bacteriophage λ particles and denatures the DNA that they contain. Place all the membranes together, DNA side up, in a plastic lunch box filled with denaturation solution. The box should be large enough to accommodate the membranes without bending or creasing and should contain sufficient liquid to allow them to move around while the box is gently shaken from side to side. The treatment need last for only one minute, after which you can remove the membranes from the box. Drain them by holding them vertically by one edge and touching the opposite edge on the side of the box. Place them, DNA side uppermost, on low-grade filter paper so that excess liquid is absorbed. This will minimize the amount of denaturation solution carried over into the neutralization solution.

◇ Denaturation solution is typically 0.4 M NaOH, 1.5 M sodium chloride.

◇ We have never had problems as a result of treating all of the membranes together in the same box.

◇ Place nylon membranes directly into the solution, but carefully float nitrocellulose filters on the surface so that they wet from one side (see Chapter 7, section 1), before immersing them.

1.3.2 Neutralization

After denaturation, place the membranes into neutralization solution, shake them gently for one minute, take them out and remove excess liquid as before.

◇ Neutralization solution is typically 1.5 M sodium chloride, 0.5 M Tris-HCl, pH 7.4.

1.3.3 Treatment with 2 × SSC

After neutralization, place the membranes into 2 × SSC for one minute, remove them and drain them as before. Place them, DNA side up, on low-grade filter paper.

◇ 2 × SSC is 300 mM sodium chloride, 30 mM sodium citrate.

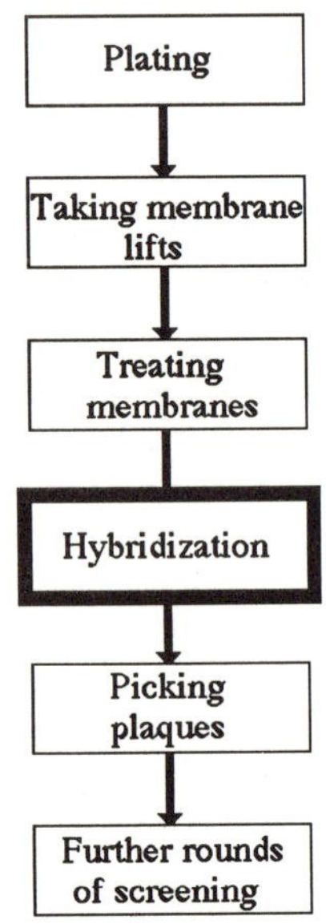

1.3.4 Fixing the DNA to the membrane

Finally, you should fix the bacteriophage λ DNA to the membrane according to the manufacturer's instructions, as discussed in Chapter 3, section 1.5. You can store membranes until you are ready to hybridize them as described in Chapter 3, section 1.6.

1.4 A brief note about hybridization probes

Hybridization and washing of the membranes should be carried out under standard, simple conditions, as discussed in the volume on hybridization in this series. It is worth just noting here that if you plan to use a cloned DNA fragment as a probe, you *must* ensure that the probe does not contain any sequences found in the bacteriophage λ vector you are using, otherwise every plaque will hybridize. This could occur, for example, if both the plasmid from which your probe was isolated and the bacteriophage λ vector contained the *E. coli lacZ* gene.

1.5 Orientating the membranes and X-ray film before autoradiography

◇ The most common error of those doing this for the first time is not to orientate membranes and film unequivocally.

◇ You may not be able to see the outline of the membranes on the exposed film. Even if you can, this will not give you accurate positioning.

◇ Do not use kitchen cling film. It generates static electricity and will expose the film.

◇ You must use unexposed X-ray film in a darkroom with an appropriate safety light.

◇ Do *not* fix radioactive dot labels to the top of the Saran Wrap, as suggested in some protocols. The Saran Wrap may shift with respect to the membranes.

After hybridizing and washing the membranes, you will be ready to autoradiograph them. When autoradiographing Benton-and-Davis membranes, it is of crucial importance to know the *exact* position of the membranes with respect to the X-ray film. If you fail to do this, you will not be able to determine exactly where on the membrane the probe has bound, and so you will not be able to determine which plaque has hybridized to the probe.

The simplest thing to do is to use the edges of the autoradiography cassette to help you to align the membranes and the film (*Figure 6.9A*). To do this, tape the membranes to the autoradiography cassette and cover them in Saran Wrap. If you leave this out, the membranes may stick to the film. Place the X-ray film in the cassette so that two adjacent edges abut two adjacent edges of the cassette. Crease one corner of the film, to remind you of its orientation in the cassette, and close the cassette.

Alternatively, you can use radioactive ink (*Figure 6.9B*). To do this, first tape the membranes to the autoradiography cassette. Mark some adhesive dot labels with spots of radioactive ink and stick them to the cassette at various asymmetric positions around the membranes. Next, cover the membranes and dot labels with Saran Wrap, place the X-ray film on top and close the cassette. In this case there is no need to make any record of the exact position of the film in the cassette.

A third method is to use carefully positioned pencil marks (*Figure 6.9C*). To do this, put a sheet of Whatman 3MM filter paper in the cassette. Wrap the membranes individually in Saran Wrap and tape

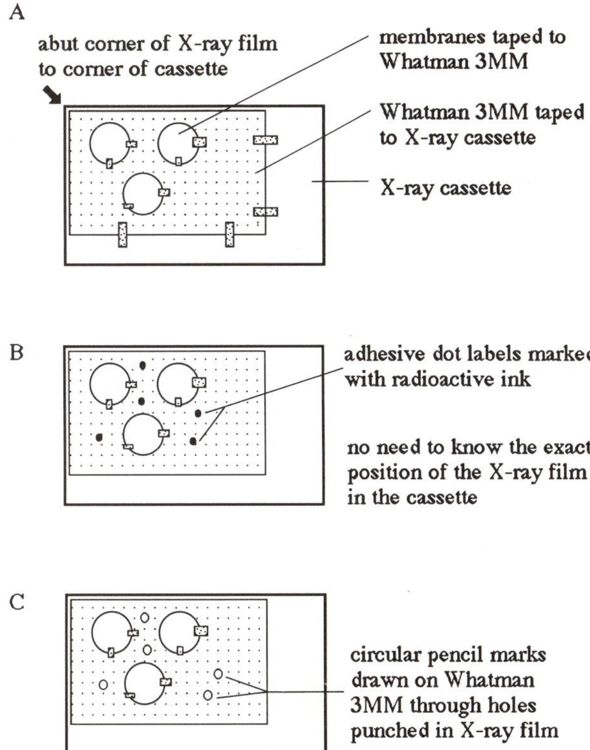

Fig 6.9

Three methods of aligning Benton-and-Davis or Grunstein–Hogness membranes with X-ray film. **A**. Using the edges of the autoradiography cassette. **B**. Using radioactive ink. **C**. Using pencil marks. Details of each method are given in the text

them to the 3 MM filter paper. Use an office hole punch to make some holes in an asymmetric pattern in the film. Put the film in the cassette and use a sharp pencil to mark the positions of the holes in the film on to the 3 MM filter paper below.

You will find that you develop a preference for one or other of these methods, or may adopt or invent some other suitable technique to achieve the same end.

1.6 Identifying hybridization signals after autoradiography

After exposing and developing the film, allow the cassette, with the membranes still attached, to warm to room temperature and wipe condensation from the Saran Wrap covering the membranes. Place the developed film in the cassette in exactly the same position as it was during exposure, by abutting the edges of the film to the edges of the cassette, by aligning the radioactive spots on the dot labels with the corresponding signals on the film, or by matching the punch holes in the film with the pencil marks on the 3 MM filter paper. Next, mark the *exact* positions of the needle holes in the membranes on to the film, using a fine-tip, water-fast marker pen (*Figure 6.10*) .

You are now in a position to identify hybridization signals. The main criterion for distinguishing a real signal from an artefact is that

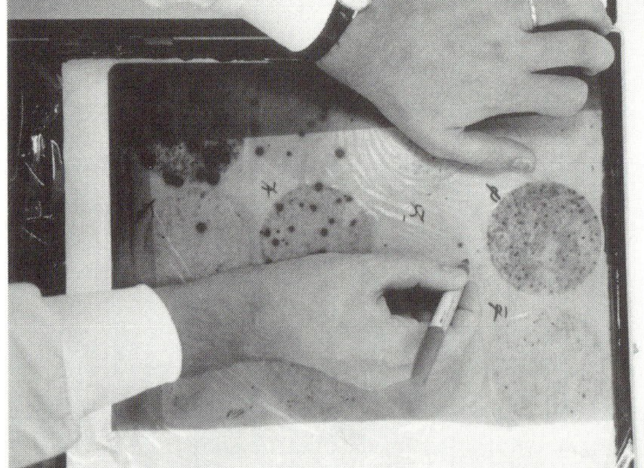

Fig 6.10

Marking the positions of the needle holes in the membranes on to the exposed and developed X-ray film, using a water-fast marker pen

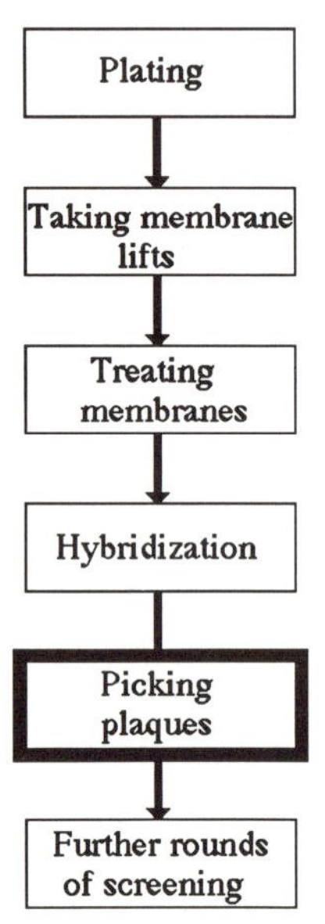

real signals appear at the same position on duplicate membranes. There may be other clues. For example, real hybridization signals frequently, though not always, have a 'comet tail' due to streaking of the bacteriophage during processing of the membranes before hybridization. With experience you will also get a feel for what 'looks real'. *Figures 6.11* and *6.12* show some examples of real signals. *Figures 6.13* and *6.14* show some of the things that can go wrong.

1.7 Picking plaques

To pick plaques, remove the plates from the refrigerator or cold-room, allow them to warm to room temperature and wipe their outsides clear of condensation. Place the marked autoradiography film on a light box and place the plate on top of the film. Remove the plate's lid and align the orientation marks on the plate with the orientation marks on the film. You should be able to see the hybridization signals through the plate. If you cannot, for example because they are too weak, mark the signals on the film with a black marker pen. When you are satisfied that everything is optimally aligned, pick the plaque(s) that lie directly above the hybridization signal.

If the plaques were plated at low density, you may be able to pick a single, well-isolated plaque by puncturing the top agarose layer with a sterile Pasteur pipette, sucking the top agarose plug up and placing it in 1 ml of SM buffer containing a drop of chloroform.

If the plaques were plated at high density you will not be able to identify the positive plaque and so will have to pick the area of top agarose that corresponds to the hybridization signal. This will contain several plaques, including the one that gave rise to the hybridization signal. As before, the plug of top agarose should be placed in 1 ml of sterile SM buffer containing a drop of chloroform.

◇ If you use cold plates, a film of condensation will form and obscure the plaques.

◇ The chloroform will lyse any viable *E. coli* that remain, maximizing the yield of bacteriophage λ and establishing bacterial sterility.

◇ To be sure of picking an area containing the correct plaque, we use the top, wide end of a sterile Pasteur pipette to pick from high density plates.

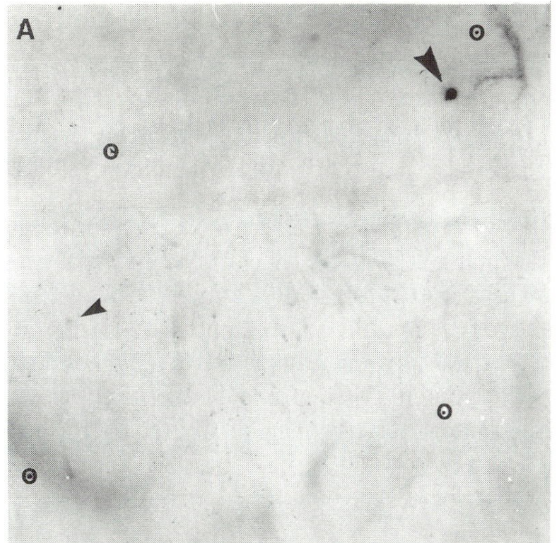

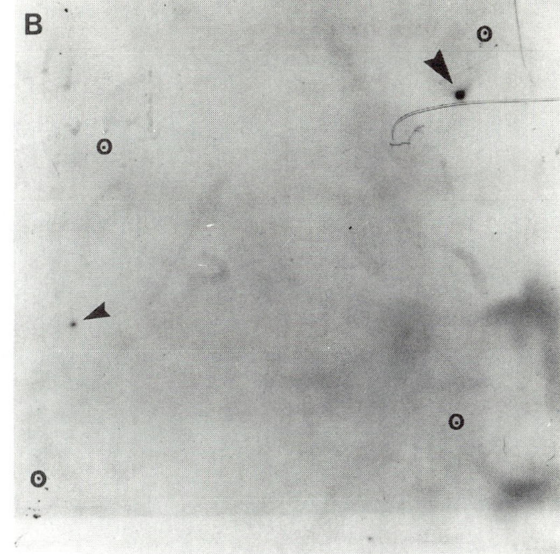

Fig 6.11

A high density Benton-and-Davis screen. 200 000 plaques of a chick embryo cDNA library, were plated on to a 20 × 20 cm dish. Duplicate membrane lifts (**A** and **B**) were hybridized with a mouse *Bmp-2* (bone morphogenetic protein-2) cDNA probe. There is a strong duplicate signal (large arrow). On **B**, this has a 'comet tail', extending to the 'north-east'. The plaque giving this signal was a chicken *Bmp-2* cDNA clone. There is also a very weak duplicate signal (small arrow). The plaque giving this signal was a cDNA clone for the related *Bmp-4* gene. Both autoradiograms have artefactual splodges, swirls and scratches that did not interfere with identification of positive plaques. Orientation marks are ringed

Fig 6.12

Two low density, Benton-and-Davis, secondary screens. 90-mm circular membranes were lifted from plates of approximately 100 plaques. Orientation marks are ringed. Many of the plaques on **A** have hybridized, giving signals with 'comet tails'. Four plaques on **B** have hybridized and there is weak non-specific hybridization to other plaques. In both cases, duplicate membranes gave identical hybridization patterns

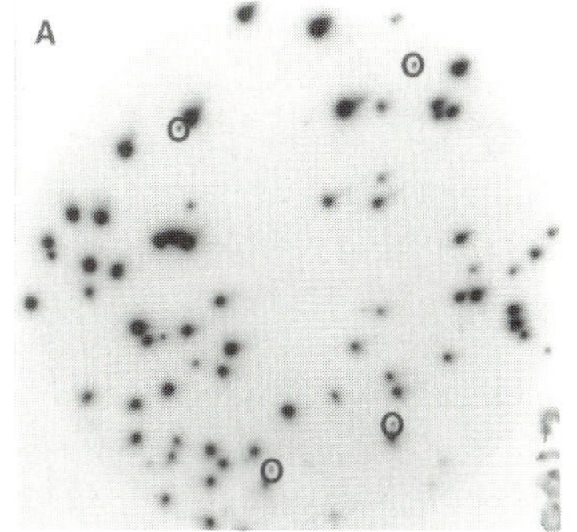

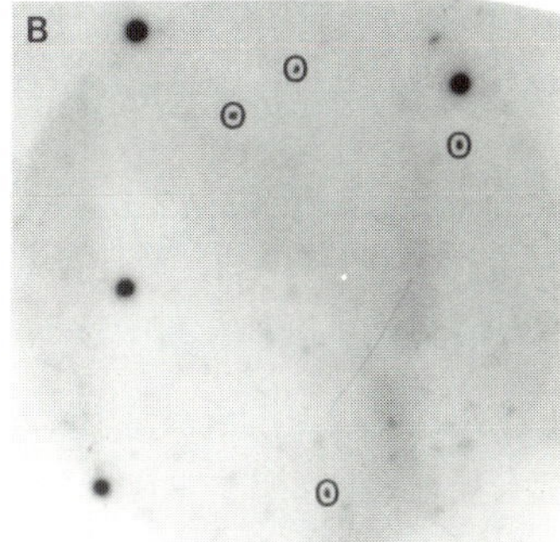

Plating

Taking membrane lifts

Treating membranes

Hybridization

Picking plaques

Further rounds of screening

1.8 Further rounds of screening

If you have picked a mixture of plaques from the first, or primary, screen you will need to plate this mixture out at low density and rescreen it for hybridizing plaques. In order to determine how much of the mixture to plate you could titre it as described in section 2. If you have a large number of mixtures of plaques to rescreen, this can be very laborious and expensive. To save time and money, you can usually assume that 50 μl of a 10^{-2} dilution of a mixture of plaques picked as described above will contain approximately 500 bacteriophage λ particles. These can be conveniently plated on a 90 mm dish. It is probably best to determine the titre of one or two of your own bacteriophage λ mixtures and then assume that the rest will have roughly similar titres.

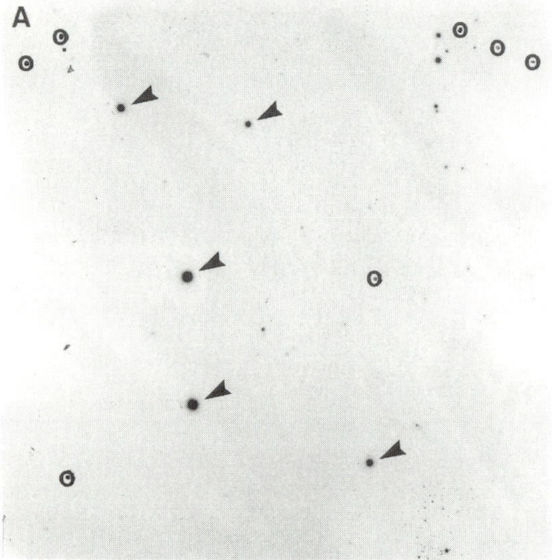

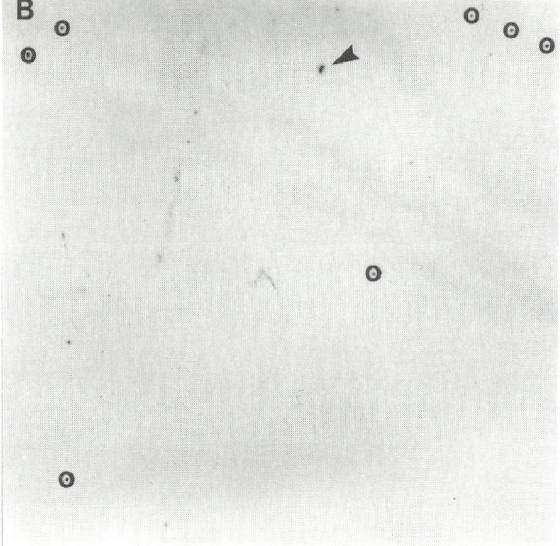

Fig 6.13

A high density Benton-and-Davis screen. The 5 strong signals on membrane **A** (arrowed) are absent from duplicate membrane, **B**. The strong signal on **B** (arrowed) is absent from **A**. To our cost, we picked plaques from all 6 regions. None contained plaques that hybridized with the probe. All 6 signals were artefacts. This underlines the importance of looking for duplicate signals. Orientation marks are ringed

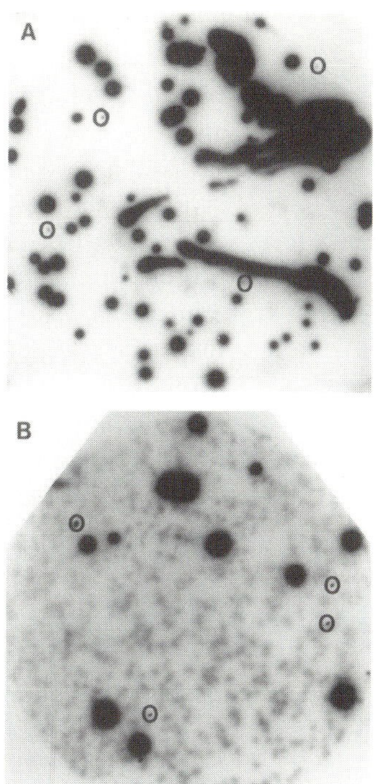

Fig 6.14

Things that can go wrong with Benton-and-Davis screening. **A**. Plaques streaked and ran into each other when condensation formed on the top agarose during plaque growth (see section 1.1). **B**. Some plaques give strong signals, but there is weak hybridization to every plaque. Exposing the X-ray film for less time, would reduce the background and real signals would be smaller, allowing positive plaques to be pin-pointed. **C**. The intensities of real signals (some are arrowed) and background signals are very similar. Picking plaques accurately might be difficult

◇ The titre will vary according to the strain of bacteriophage λ, the size of the top agarose plug you picked, and other factors.

◇ This will help you to avoid errors resulting from growing up and analysing a mixed population of clones.

It may happen that after a secondary screen you are still unable to pick a single, well-isolated hybridizing plaque, and have to pick a plaque that is contaminated with one or two neighbours. In this case, you will have to perform a tertiary screen, perhaps with only 50 to 100 plaques on a 90 mm dish. In any case, once you have picked what you believe to be a pure plaque, it pays to re-plate an aliquot of it at low density and to check that every plaque now hybridizes.

2. Determining the titre of a bacteriophage λ suspension

The obvious way to determine the titre of a bacteriophage λ suspension is to make a series of dilutions of the suspension in SM buffer and to plate aliquots of the diluted samples on a series of 90 mm dishes as described in section 1.1.

A quicker, although slightly less accurate, method is to dispense aliquots of the diluted bacteriophage suspensions on to a spot plate. To do this, you need to pour a base agar plate and then a top agarose layer that contains *E. coli* but no bacteriophage. When the top agarose has set completely, draw a grid of numbered squares on the bottom of the dish with a water-fast marker pen. Spot 5 μl of each diluted bacteriophage suspension on to the surface of the plate, so that each spot falls within a numbered grid square. Leave the plate for approximately 30 minutes to allow the liquid in the spots to soak into the surface of the top agarose and then incubate the plate, inverted, overnight. The next day, you will find that plaques have developed from each spot. If you count the number of plaques that have developed from each dilution, you will be able to calculate the approximate titre of the original suspension.

◇ Take care that the spots are spaced sufficiently far apart so that they do not run into each other.

3. Screening bacterial colonies by the Grunstein–Hogness method

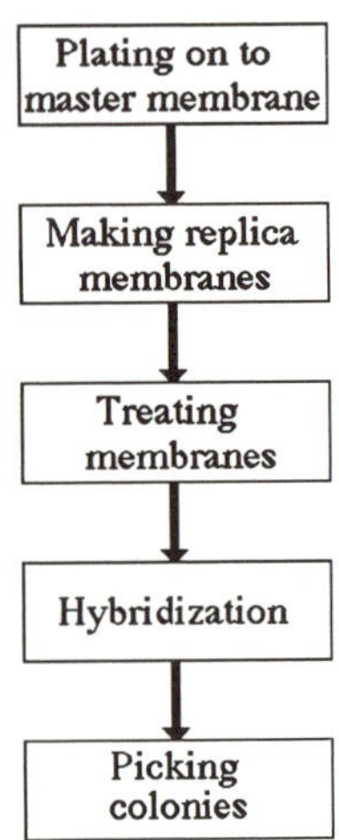

The most commonly used method for screening collections of bacterial colonies containing recombinant plasmids is the one developed by Grunstein and Hogness (1975). You can use this method to search through small collections of recombinants, such as the products of a sub-cloning experiment, or to screen whole cDNA or genomic libraries constructed in plasmid or cosmid vectors.

The steps you have to take in the Grunstein–Hogness procedure (*Figure 6.15*) are as follows:

(1) Plate the collection of bacterial colonies out on a nylon membrane or nitrocellulose filter lying on top of an agar layer in a Petri dish. This is the *master membrane*.
(2) Incubate the plate overnight to allow bacterial colonies to grow.
(3) Make a replica of the master membrane by removing it from the surface of the agar and placing it in close contact with a new membrane, so that some bacteria from each colony are transferred to the new membrane.
(4) Peel the two membranes apart and place them, bacteria uppermost, on fresh agar plates.
(5) After a further period of growth, make a second replica of the master membrane on a second new membrane. Incubate both of these membranes on fresh agar plates. After this period of growth, store the master membrane and screen the two replica membranes.

◇ The proliferation of agar plates that this procedure entails makes it more awkward to perform than Benton-and-Davis screening.

How do you screen the membranes? First, the membranes are treated to release DNA from the *E. coli* and to fix it to the membrane. Specific DNA sequences on the membrane are then detected by hybridization with a labelled probe, followed by autoradiography (or the appropriate method for detecting a non-radiocative probe).

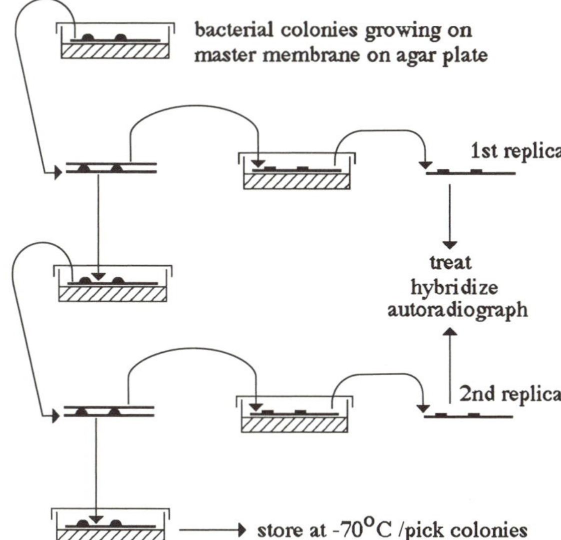

Fig 6.15

Grunstein–Hogness screening

Hybridizing colonies are identified by aligning the autoradiogram with the master membrane and picked for further analysis.

We will now look at some details of this procedure.

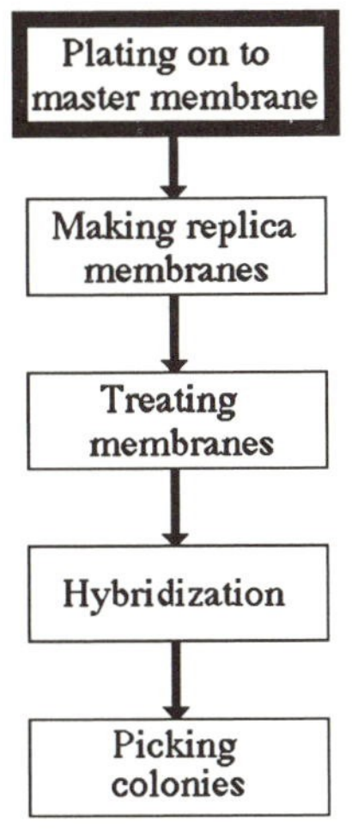

3.1 Plating

As with Benton-and-Davis screening, you must first decide which size of Petri dish to use and how many colonies to plate on each dish. The issues discussed in section 1.1.1 apply. Prepare plates containing 1.1 per cent (w/v) agar in L broth as discussed in section 1.1.2. It is not necessary to include magnesium ions or maltose in the plates, but remember to include the appropriate antibiotic to select for *E. coli* colonies containing the plasmid/cosmid that you are working with.

Plates can be used soon after they have set and there is no need to go to great lengths to dry them.

◇ Wear gloves when you handle membranes.

◇ If the manufacturer recommends that a particular side of the membrane is used to bind DNA, this side should be *uppermost*.

When you are ready to plate out the bacteria, take a fresh nylon membrane, label it with a soft lead pencil and gently lower it on to the surface of the base agar. As discussed in section 1.2, and in Chapter 7, nylon membranes are much preferred to nitrocellulose filters for this purpose. To ensure that the membrane lies flat on the surface, without trapping any bubbles of air, follow the tips given in section 1.2.

A bacterial suspension, such as a cDNA library constructed in a plasmid vector, a genomic library constructed in a cosmid vector, or a small collection of cloned fragments, can be spread on top of the membrane in exactly the same way as it would be spread on to an agar surface. Alternatively, if you have a few purified clones that you wish to test for homology to a particular probe, you can streak individual clones out in an array on the surface of the membrane, in exactly the same way as you would streak out a bacterial suspension

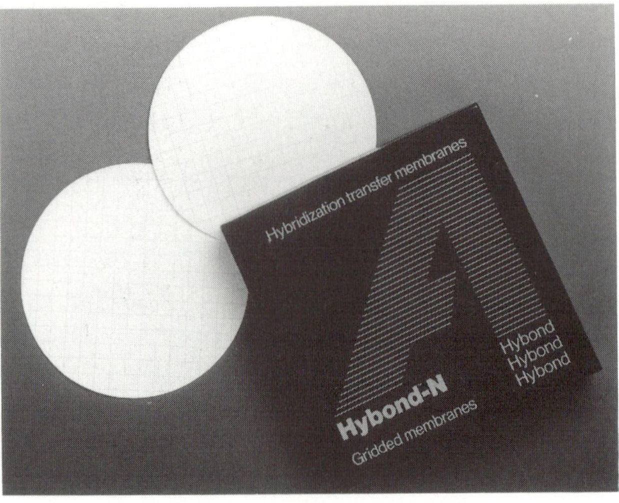

Fig 6.16

Commercially available 90 mm circular membranes with grids marked on them. (Photograph courtesy of Amersham International)

on an agar surface. Some manufacturers supply membranes with grids marked on them (*Figure 6.16*), so that you can streak out each clone in an easily identifiable grid square. Alternatively, you can draw your own grids on to membranes using a pencil or a ball-point pen.

Since non-specific background hybridization to bacterial colonies can be a problem, include a positive and negative control if possible. For example, as a positive control you could use *E. coli* containing the recombinant plasmid from which the probe is derived, and as a negative control you could use *E. coli* containing plasmid vector.

After plating, incubate the dishes inverted at 37 °C overnight, to allow colonies to develop.

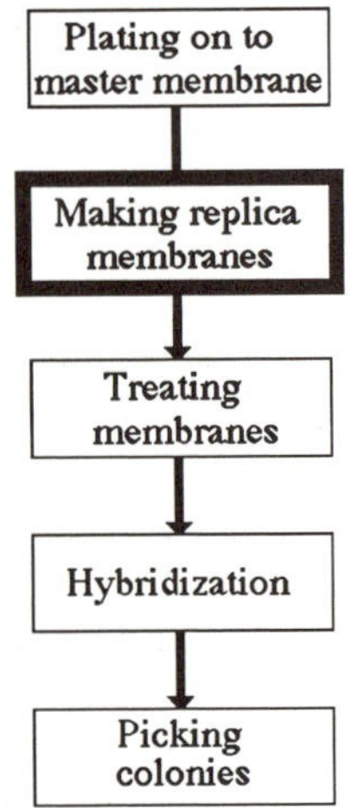

3.2 Making replica membranes

3.2.1 Making the first replica

First, carefully peel the master membrane off the plate and place it on a sheet of cling film, colonies uppermost. Label a fresh membrane with a soft lead pencil and gently lay it down on top of the master membrane. Avoid movement of the membranes with respect to each other once initial contact between them has been made. Next, press the membranes together firmly to ensure that efficient transfer of bacteria occurs. Some laboratories have a fancy, purpose-made velvet replica plating tool for this purpose. We place a sheet of cling film on top of the two membranes, place a glass plate on top of the cling film, and press firmly downwards.

◇ If one side of the membrane binds DNA preferentially it should face downwards, in contact with the colonies on the master membrane.

Orientate the membranes with respect to each other by piercing them in a number of places, in an asymmetrical pattern, with an 18 gauge needle (*Figure 6.17*). Finally, peel the membranes apart carefully and lie them, colonies uppermost, on two fresh base agar plates containing the appropriate antibiotic.

◇ Lie the membranes on a sheet of polystyrene so that the needle penetrates easily (*Figure 6.17*).

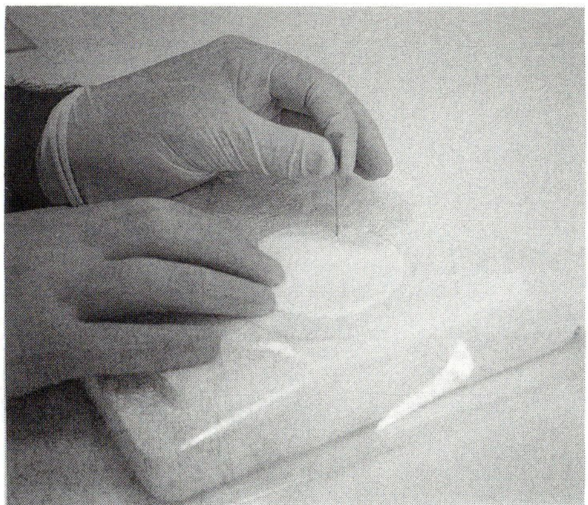

Fig 6.17

Making orientation marks in Grunstein–Hogness membranes, using an 18 gauge syringe needle. The two sandwiched membranes are placed on a clean sheet of cling film, on top of polystyrene

3.2.2 Making the second replica

Immediately after you have made the first replica membrane, the colonies on the master membrane will look rather flat. After a few hours of incubation at 37 °C, the colonies will have grown a little and will have a more rounded appearance. At this point, use the master membrane to make a second replica membrane in the same way as before. Make orientation holes in the second replica membrane at the same positions as the existing holes in the master membrane. When you have peeled the master membrane and second replica membrane apart, lie them on two fresh base agar plates containing the appropriate antibiotic and incubate for a few more hours at 37 °C.

Once colonies become visible on the two replica membranes, they can be used for screening as described in section 3.3.

3.2.3 What to do with the master membrane while you are screening the replicas

If you plan to pick colonies within a few days, leave the master membrane on the last base agar plate, wrap the plate in cling film or parafilm and store it inverted in a refrigerator or cold-room.

If you want to keep the master membrane for an extended period because, for example, you wish to re-screen the replica membranes with a series of probes or because the master membrane carries a precious library, then you must store it at –70 °C to preserve the viability of the bacteria. In this case, place the master membrane, colonies uppermost, on a fresh agar plate containing the appropriate antibiotic and 25 per cent (v/v) glycerol and incubate at 37 °C for 1 hour. Seal the plate with parafilm, seal this in a plastic bag and store the whole thing, inverted, at –70 °C. Alternatively, remove the master membrane from the plate, sandwich it with a new membrane, place the two membranes in the middle of four sheets of Whatman 3MM filter paper soaked in L broth/25 per cent (v/v) glycerol, seal this in a plastic bag, and store at – 70 °C.

◇ Bacteria remain viable indefinitely at –70 °C in the presence of glycerol.

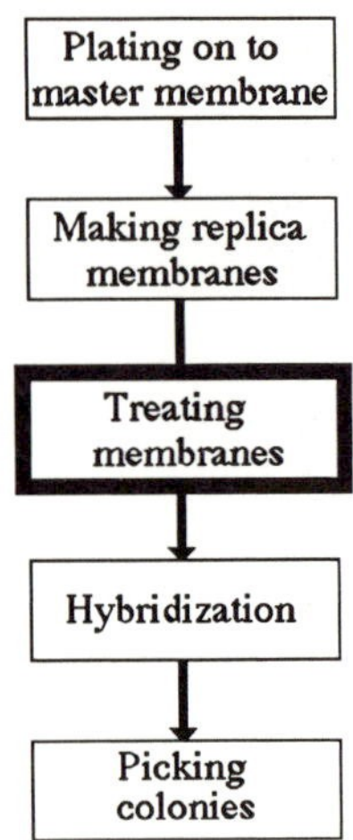

When you want to use the master membrane again, thaw it at room temperature. Colonies can then be picked from it, or new replica membranes made as described above.

3.3 Treating membranes before hybridization

The membranes are treated in turn with denaturation solution, neutralization solution, and 2 × SSC, and the plasmid/cosmid DNA is then fixed to them. The procedure is very similar to that used in Benton-and-Davis screening (see section 1.3), but you *must* take care not to wet the colony-side of the membrane during the denaturation and neutralization. If you do, the colonies will tend to run into each other and this will result in diffuse, weak, streaky hybridization signals.

The denaturation step disrupts the *E. coli* cells and denatures the DNA that they contain. To do this, lay a piece of cling film on the bench, and make a puddle of denaturation solution in the middle (*Figure 6.18A*). Lower the membrane on to the puddle, so that its underside becomes completely wet, with no air bubbles, whilst its upper, colony-bearing side remains free of liquid (*Figure 6.18*). After two minutes of wetting, carefully lift the membrane and place it, colonies uppermost, on a sheet of dry filter paper, so that it blots dry. Finally, repeat the procedure with a fresh puddle of denaturation solution.

◇ Denaturation solution is typically 0.4 M NaOH, 1.5 M sodium chloride.

◇ If the puddle is too small, the membrane will not get completely wet. If it is too large it will flood the top of the membrane. We use approximately 0.5 ml for a 90 mm circular membrane, and proportionately more for larger membranes.

After the denaturation step, you should neutralize the membrane by repeating the above procedure twice with fresh puddles of neutralization solution.

◇ Neutralization solution is typically 1.5 M sodium chloride, 0.5 M Tris-HCl, pH 7.4.

Finally, soak the membrane in 2 × SSC. By this stage, the membrane can be immersed in liquid and so we place it in a bath of 2 × SSC. While it is in the bath, gently wipe its surface with a gloved finger or with tissue paper to remove the residue of the bacterial colonies. This does not lead to a significant decrease in the quality of the hybridization signals, but does significantly reduce the level of background binding of the probe. After a thorough rinsing in 2 × SSC to remove all bacterial debris (and bits of tissue paper!), fix the DNA to the membrane as discussed in Chapter 3, section 1.5.

◇ Surprisingly, this does not seem to lead to smearing of the DNA.

◇ Some people include SDS and proteinase K in this rinse, but this is not necessary.

Membranes can be hybridized immediately, or stored as described in Chapter 3, section 1.6.

3.4 Another brief note about hybridization probes

As with Benton-and-Davis screening, it is important to ensure that your probe does not contain sequences that are homologous to the vector in which your library is constructed. If your probe is a DNA fragment that has been excised by restriction digestion from a plasmid vector, the vector is likely to share extensive regions of homology

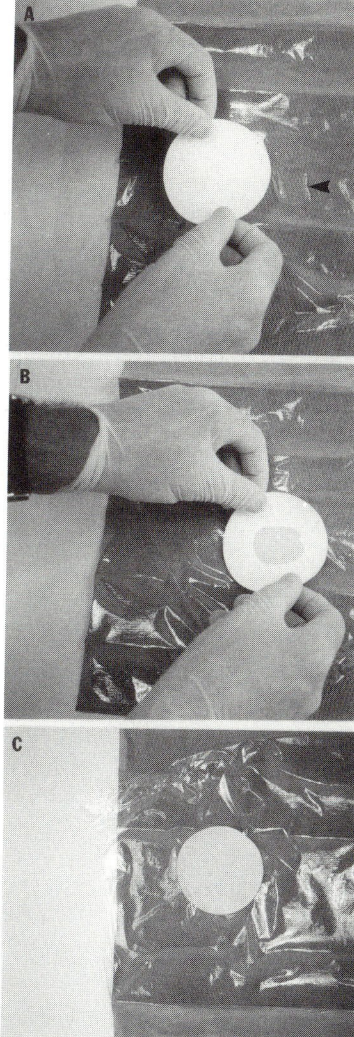

Fig 6.18

Treating Grunstein–Hogness membranes with denaturation and neutralization solutions. Do not allow the solutions to flow on to the top of the membrane. **A**. Lower the membrane on to a puddle of solution (arrowed) on cling film. **B**. Allow the membrane to wet from the centre outwards. **C**. Leave the membrane on the puddle for 2 minutes before carefully lifting it up

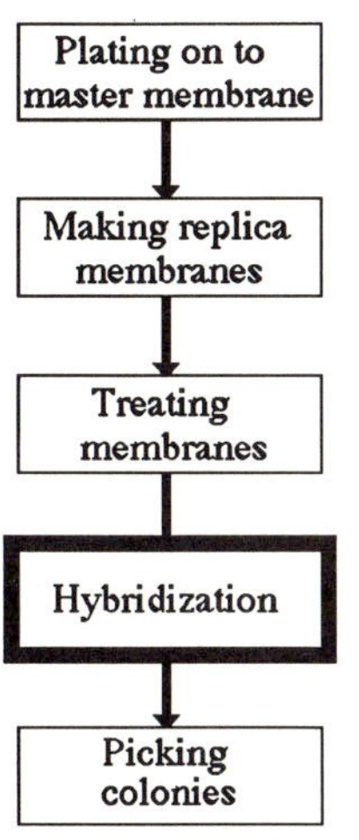

with the vector DNA on the membranes, and you must take particular care to minimize contamination of the probe with the vector. We routinely perform two rounds of purification of the insert on agarose gels. If the insert and the vector are of similar sizes it may be difficult to purify the insert without contamination. In such cases, consider cutting the insert into smaller fragments that can be easily separated from the vector by electrophoresis. You could then label one, or a mixture, of these fragments for use as your probe. It is worth taking the time to hybridize the purified, labelled probe to a Southern blot of the digested plasmid from which it was derived, to check that there is no significant background hybridization to vector sequences.

3.5 Orientating the membranes and X-ray film before autoradiography

As with Benton-and-Davis screening, it is of crucial importance that you know the exact position of the membranes with respect to the X-ray film, using one of the techniques discussed in section 1.5.

After autoradiography, you will be able to look for hybridization signals. Once again, the main criterion to be used in judging whether a signal is real is that a duplicate signal is found at the same position on the duplicate membrane. Signals of hybridization to bacterial colonies sometimes have a 'comet tail', but this is less diagnostic than for signals of hybridization to bacteriophage λ plaques. Examples of hybridization signals obtained by Grunstein–Hogness screening are shown in *Figures 6.19*, *6.20*, and *6.21*.

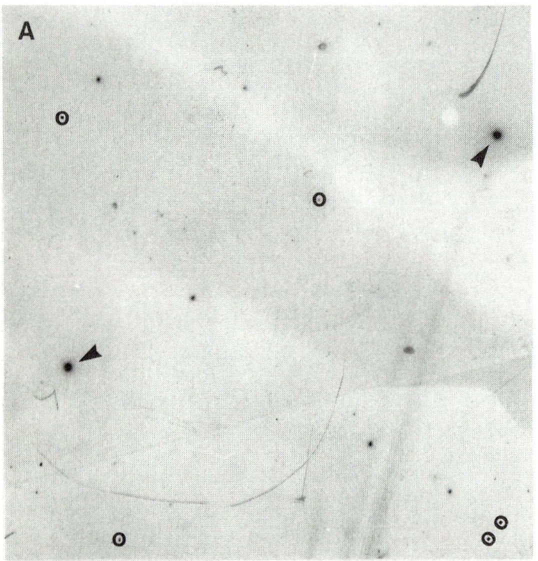

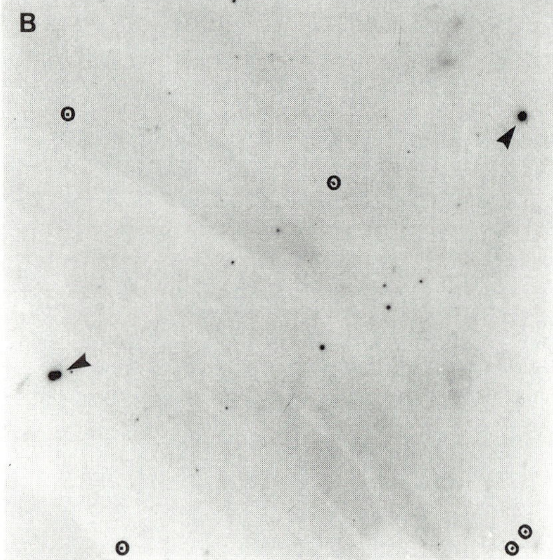

Fig 6.19

A high density Grunstein–Hogness screen. 250 000 cosmid genomic clones were plated on a 20 × 20 cm dish. Replica membranes (**A** and **B**) were hybridized with a c-*fgr* cDNA probe. The two strong duplicate signals (arrowed) were shown to correspond to c-*fgr* genomic clones (Patel *et al.* 1990). The weaker signals are not duplicated on both membranes. Orientation marks are ringed

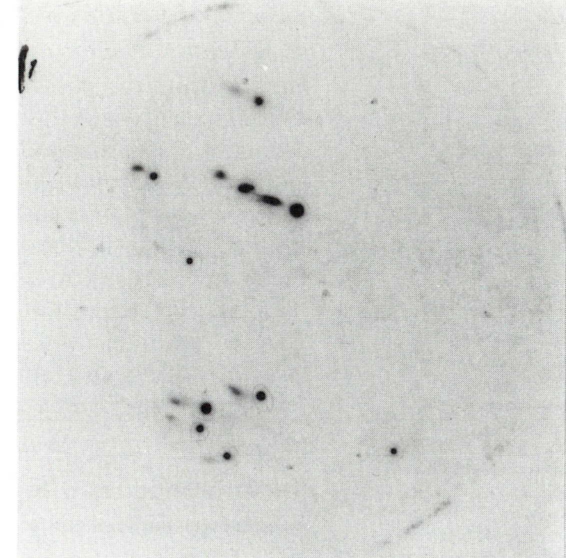

Fig 6.20

A low density Grunstein–Hogness screen. There are several strong signals. All were present on a duplicate membrane. There is very weak hybridization to the other colonies

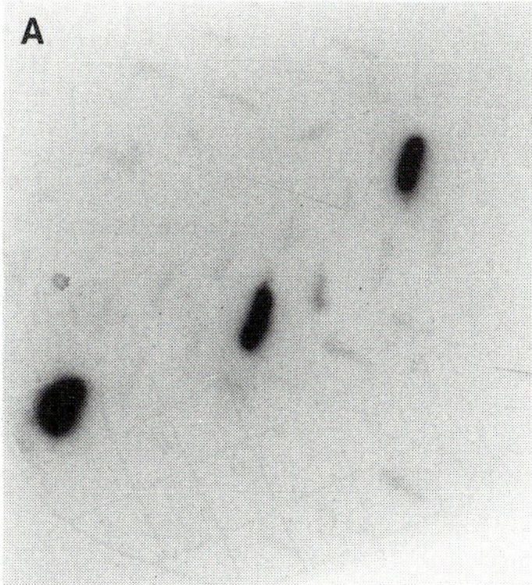

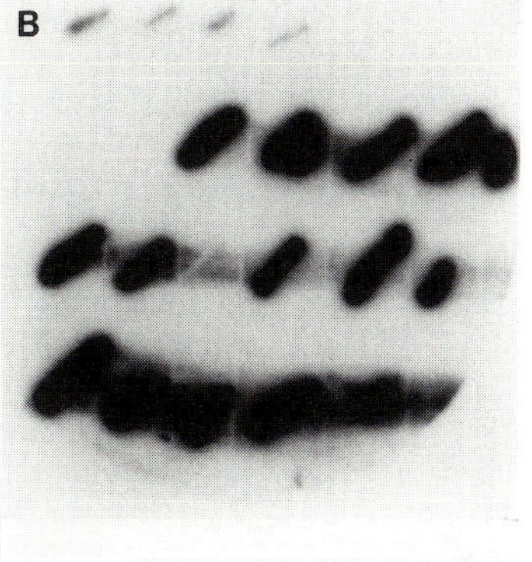

Fig 6.21

Grunstein–Hogness screen of membranes on which bacterial colonies were streaked in grids. Hybridizing and non-hybridizing colonies are easily distinguishable, even on **B**, where colonies have run into each other and the film is over-exposed. There are no orientation marks on the membranes, but the disposition of colonies was asymmetrical, so that the autoradiographs could be matched to the master membrane

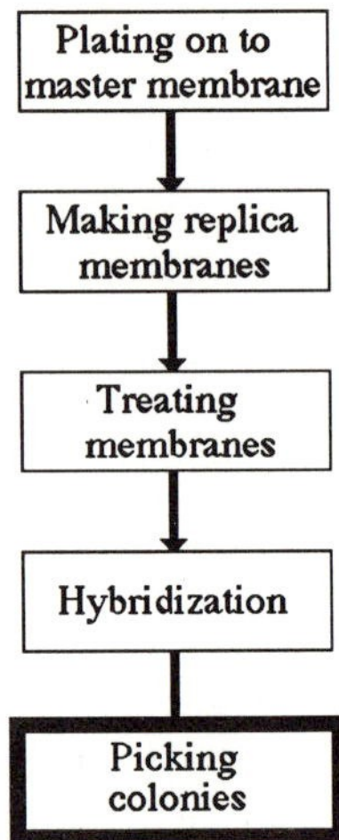

◇ If the signals are too weak to be seen through the membrane, mark their positions on the film with a black marker pen.

3.6 Picking colonies

Having identified a positive signal on an X-ray film, place the film on a light box and cover it with a sheet of cling film. Retrieve the master membrane, place it on top of the cling film and align the orientation marks on the membrane with those on the film. You should be able to see hybridization signals through the membrane. When you are satisfied that the membrane and the film are optimally aligned, pick the colony/colonies that lie directly above the hybridization signal by scraping them off with a sterile toothpick and transferring them to 1 ml of L broth in a small tube. If you cannot pick a single colony, you will have to perform further rounds of Grunstein–Hogness screening, taking note of the comments made in section 1.8.

4. Further reading

Berger, S.L. and Kimmel, A.R. (ed.) (1987). Selection of clones from libraries. *Methods in Enzymology*, **152**, 393–504.

Perbal, B. (1988). *A practical guide to molecular cloning*, (2nd edn), Wiley, New York, pp. 510–15; 421–3.

Sambrook, J., Fritsch, E.F., and Maniatis, T. (1989). *Molecular cloning: a laboratory manual*, (2nd edn), Vol. 1, pp. 2.3–2.63; 2.108–2.121; 1.90–1.104. Cold Spring Harbor Laboratory Press.

5. References

Benton, W.D. and Davis, R.W. (1977). Screening λgt recombinant clones by hybridization to single plaques in situ. *Science*, **196**, 180–2.

Cowell, I.G. and Hurst, H.C. (1993). Cloning transcription factors from a cDNA expression library. In *Transcription factors: A practical approach*, (ed. D.S. Latchman), pp. 105–23, IRL Press at Oxford University Press, Oxford.

Frischauf, A-M., Lehrach, H., Polstka, A., and Murray, N.M. (1983). Lambda replacement vectors carrying polylinker sequences. *Journal of Molecular Biology*, **170**, 827–42.

Grunstein, M. and Hogness, D.S. (1975). Colony hybridization: a method for the isolation of cloned DNAs that contain a specific gene. *Proceedings of the National Academy of Sciences, USA*, **72**, 3961–5

Helfman, D.M. and Hughes, S.H. (1987). Use of antibodies to screen cDNA expression libraries prepared in plasmid vectors. *Methods in Enzymology*, **152**, 451–7.

Huynh, T.V., Young, R.A., and Davis, R.W. (1984). Constructing and screening cDNA libraries in λgt10 and λgt11. In *DNA cloning: A practical approach*, (ed. D.M. Glover), pp. 49–78, IRL Press at Oxford University Press.

Mierendorf, R.C., Percy, C., and Young, R.A. (1987). Gene isolation by screening λgt11 libraries with antibodies. *Methods in Enzymology*, **152**, 458–69.

Patel, M., Leevers, S., and Brickell, P.M. (1990). Structure of the complete human c-*fgr* proto-oncogene and identification of multiple transcriptional start sites. *Oncogene*, **5**, 201–6.

Short, J.M. and Sorge, J.A. (1992). *In vivo* excision properties of bacteriophage λZAP expression vectors. *Methods in Enzymology*, **216**, 495–508.

7 Filters and membranes

When Southern blotting was first developed, DNA was transferred to nitrocellulose filters (Southern 1975). RNA had been found not to bind efficiently to nitrocellulose filters and so, in the first northern blots, RNA was transferred to DBM paper (Alwine *et al.* 1977), where the nucleic acids bound covalently to reactive diazo groups (see Chapter 1). In fact, following this, DBM paper became quite widely used for Southern blotting, since it was more efficient at retaining short DNA molecules than were nitrocellulose filters. However, the method used for diazotization was rather cumbersome and, since the reactive groups had a short half-life, the diazotized paper had to be prepared immediately before use. The situation improved when *o*-aminophenylthioether-coupled filter paper (APT paper) was developed (Seed 1982), since this was easier to prepare and was more stable than DBM paper. However, the use of derivatized filter paper was largely discontinued when nitrocellulose filters that would bind DNA and RNA with high efficiency became available.

There are a number of problems associated with the use of nitrocellulose filters, as discussed below. To overcome these, a number of commercial suppliers developed membranes based on nylon. Some of the most commonly used nitrocellulose filters and nylon membranes are listed in *Table 7.1*. What are the differences between these products?

- Both nitrocellulose filters and nylon membranes are available in *unsupported* forms, where the active substrate is present as a pure cast sheet, and *supported* forms, where the active substrate is cast on to a sheet of inert material
- Nylon membranes may have an *uncharged* or a *positively charged* surface. Some, although not all, manufacturers claim that their charged membranes have a greater nucleic acid binding capacity than their uncharged membranes
- Nylon membranes from different companies differ in the structure of the nylon weave and in the method used to apply a charge to them.

Table 1.1 Some commercially available nylon membranes and nitrocellulose filters that have been designed for blotting. The list is not exhaustive and is in alphabetical order, rather than in order of preference. Some details are unavailable. Some manufacturers offer membranes in a range of pore sizes and recommend particular applications for each product. Note that GeneScreen*Plus* was once manufactured so that one side bound nucleic acids with much higher efficiency than the other. This is no longer the case, and both sides work equally well

Name	Type	Supplier
Biodyne	nylon	Pall Ultrafine Filtration Corporation
Biotrans	uncharged nylon supported	ICN Biomedicals Inc
Biotrans+	positively charged nylon supported	ICN Biomedicals Inc
Duralon-UV	uncharged nylon	Stratagene Cloning Systems
Duralose-UV	nitrocellulose supported	Stratagene Cloning Systems
GeneBind	positively charged nylon	Pharmacia Biotech Limited
GeneScreen	uncharged nylon supported	Du Pont (UK) Limited
GeneScreen*Plus*	positively charged nylon supported	Du Pont (UK) Limited
Hybond-N	uncharged nylon supported	Amersham International plc
Hybond-N+	positively charged nylon supported	Amersham International plc
Hybond-C	nitrocellulose unsupported	Amersham International plc
Hybond-C extra	nitrocellulose supported	Amersham International plc
Hybond-C super	nitrocellulose supported optimized for western blotting	Amersham International plc
Hybond-ECL	nitrocellulose unsupported optimized for ECL detection	Amersham International plc
OptiBLOT	positively charged nylon	International Biotechnologies
Zeta-Probe	positively charged nylon	Bio-Rad Laboratories Limited

1. The advantages and disadvantages of nitrocellulose filters and nylon membranes?

The early nylon membranes, developed in the mid-1980s, were of variable quality and, as late as 1987, lab manuals were stating that nitrocellulose filters were to be preferred for Southern and northern blotting. However, the nylon membranes that are currently available are superior to nitrocellulose filters in many respects and we recommend *without reservation* that you use them. We will now discuss the reasons for this.

1.1 Nylon membranes are physically strong

Nylon membranes have great physical strength. You will find it almost impossible to tear them. In contrast, nitrocellulose filters are fragile and easily torn, and become even more brittle after treatment with alkali (as used in colony hybridization experiments), or after incubation at the high temperatures used for hybridization. Not only does this make nitrocellulose filters harder to work with, it also means that they can only reliably be hybridized once. In contrast, nylon membranes can be hybridized sequentially with a number of different probes without falling to pieces. This is particularly useful if your membrane is carrying precious nucleic acid samples, and also saves time.

1.2 DNA and RNA bind covalently to nylon membranes

The nature of the interaction between nucleic acids and nitrocellulose is not understood, but is assumed to be non-covalent. In contrast, nucleic acids can be covalently cross-linked to nylon membranes by UV irradiation or, in some cases, simply by drying. This means that nucleic acids will remain attached to a nylon membrane through many sequential rounds of hybridization.

1.3 Nylon membranes have a high nucleic acid binding capacity

Nylon membranes have a greater nucleic acid binding capacity than nitrocellulose filters. Typical binding capacities are 480–600 μg DNA/cm^2 for the nylon membrane Hybond-N and 80–100 μg/cm^2 for the nitrocellulose filter Hybond-C (*Table 7.1*).

1.4 Nylon membranes are hydrophilic

Nitrocellulose filters are hydrophobic. Before they are used they must be carefully floated on water so that they become thoroughly wetted by capillary action. If you fail to do this, or if you plunge the dry filter into water, air becomes trapped within its pores, resulting in inefficient and uneven transfer. Do not try to wet nitrocellulose filters by floating them on a transfer solution such as 20 × SSC, because such solutions do not wet nitrocellulose filters efficiently. When performing colony or plaque lifts (see Chapter 6), nitrocellulose filters should be used dry, since fluid carried over by wet filters will cause the colonies/plaques to run into each other. Nitrocellulose filters must therefore be lowered very cautiously on to agarose plates, so that they become thoroughly and evenly wetted by water in the top agarose. Moreover, nitrocellulose filters have a tendency to slip once they have been placed on the surface of the top agarose, again as a result of their hydrophobicity. This will reduce the accuracy with which hybridizing colonies/plaques can be matched back to the bacterial plate. In contrast, nylon membranes are hydrophilic. They do not need to be wetted before use and are very easy to handle in colony/plaque lifting experiments.

1.5 Nylon membranes retain their size and shape at high temperatures

Nitrocellulose filters have a tendency to become slightly distorted when incubated at high temperatures, as during hybridization and washing. This can make it difficult to locate accurately the hybridizing plaques or bacterial colonies in a library screen. Nylon membranes are not subject to such distortion.

1.6 Nylon membranes are not inflammable

Nitrocellulose filters are highly inflammable. If you intend to hybridize nitrocellulose filters in a plastic bag sealed with a heat sealer, you must be very careful not to allow the heated strip to touch the filters. One of us did this accidentally on one occasion. A stack of 10 precious filters ignited with a brilliant orange flash and vapourized, leaving an apparently empty plastic bag. This was spectacular, but deeply upsetting and potentially dangerous. Do not do it. As far as we know, nylon membranes are not inflammable under normal conditions of use.

1.7 Nylon membranes do *not* require solutions of high ionic strength to bind nucleic acids efficiently

Nitrocellulose filters only bind nucleic acids in solutions of high ionic strength, whilst nylon membranes will bind nucleic acids equally well in solutions of high and low ionic strength. For reasons discussed in Chapter 3, section 4.1, this means that nylon membranes are suitable for electroblotting, whilst nitrocellulose filters are not.

1.8 Nylon membranes may give higher backgrounds than nitrocellulose filters, but this can be overcome easily

The only important disadvantage of the use of nylon membranes is that they tend to bind higher background levels of probe than do nitrocellulose filters. This is particularly noticeable with RNA probes. However, this has become less of a problem as the construction of nylon membranes has been refined over the years, and high background signals can be suppressed by careful use of blocking agents during prehybridization and hybridization.

1.9 Nylon membranes are sometimes 'single-sided'

A trivial problem associated with the use of some, although not all, nylon membranes is that their two surfaces do not bind nucleic acids equally well. You should be careful to follow the manufacturer's instructions in identifying the correct side to place next to your gel or bacterial plate.

1.10 Nylon membranes and nitrocellulose filters must be handled with care

Both nylon membranes and nitrocellulose filters must be handled with care. Grease from your fingers and particles of dirt can reduce the efficiency with which nucleic acids transfer and bind to the membrane/filter and can increase background binding of the probe during hybridization. *Always* wear disposable gloves when handling membranes/filters. When cutting and preparing membranes/filters before blotting, *always* keep them between the protective sheets provided by the manufacturer. Some people manœuvre membranes/filters with blunt forceps, but we prefer to hold them at the edges with gloved fingers.

2. Which membrane should be used?

It should be apparent from the above that we strongly advise you to use nylon membranes rather than nitrocellulose filters for all of the procedures discussed in this book. Or were we, perhaps, too subtle? In particular, nitrocellulose filters should *not* be used when screening libraries by hybridization. We only use nitrocellulose filters for western blotting of proteins or for screening libraries with antisera, which will be discussed in another volume in this series.

The choice of which nylon membrane to use is largely a matter of personal preference. We have used all of the membranes listed in *Table 7.1*, with good results. Stacey and Jakobsen (1993) have made a controlled study of some of the most commonly used nylon membranes, to determine which performed best in DNA-fingerprinting applications.

Whichever membrane you choose, we recommend that you follow closely the instructions provided by the manufacturer for blotting, hybridization, and washing.

3. References

Alwine, J.C., Kemp, D.J., and Stark, G.R. (1977). Method for detection of specific RNAs in agarose gels by transfer to diazobenzyloxymethyl-paper and hybridization with DNA probes. *Proceedings of the National Academy of Sciences, USA*, **74**, 5350–4.

Seed, B. (1982). Diazotizable arylamine cellulose paper—the coupling and hybridization of nucleic acids. *Nucleic Acids Research*, **10**, 1799–1810.

Southern, E.M. (1975). Detection of specific sequences among DNA fragments separated by gel electrophoresis. *Journal of Molecular Biology*, **98**, 503–17.

Stacey, J.E. and Jakobsen, K.S. (1993). Testing of nylon membranes for DNA-fingerprinting with multilocus probes. *International Journal of Genome Research*, **2**, 159–65.

Glossary

Agar A polymer extracted from seaweed. Used as a matrix on which to grow *E. coli* and other microorganisms.

Agarose A polymer extracted from seaweed. More pure than agar. Used as a matrix for electrophoresis gels.

Autoradiography A technique for detecting radiolabelled molecules by exposure of X-ray sensitive photographic film.

Bacteriophage λ A virus that infects *E. coli*. Typically used for constructing libraries of moderate-sized (15–20 kb) genomic fragments.

Benton-and-Davis screening A technique for detecting recombinant bacteriophage λ that carry a particular insert, by means of hybridization with a labelled probe.

Capillary blotting A technique for performing Southern or northern blotting, in which capillary action is used to draw solution through a gel, carrying nucleic acids out of the gel, and on to a membrane, in the flow of solution.

Clone A population of genetically identical cells or molecules, such as *E. coli* containing identical recombinant plasmids, or bacteriophage λ containing identical inserted foreign DNA sequences.

Complementary nucleic acid sequences Two nucleic acid strands that can form a double-stranded molecule by hydrogen bonding between their base pairs.

Complementary DNA (cDNA) library A collection of clones with inserts corresponding to each mRNA in a cell or tissue.

Cosmid A plasmid vector that contains the bacteriophage λ *cos* site. Typically used for constructing libraries of large (35–40 kb) genomic fragments.

Denaturation of nucleic acids Separation of two complementary strands of nucleic acid by disruption of the hydrogen bonds that hold the base pairs together.

Deoxyribonuclease (DNase) An enzyme that degrades DNA.

Depurination A technique for breaking large DNA fragments into smaller fragments, after gel electrophoresis, to facilitate their transfer from the gel to a membrane.

Dot blot A technique for applying nucleic acids to tightly restricted areas of a membrane, in the shape of small circular dots.

Duplex A double-stranded nucleic acid molecule.

Electroblotting
A technique for performing Southern or northern blotting, in which nucleic acids are induced to migrate out of a gel and on to a membrane, under the influence of an electric field.

Ethanol precipitation Precipitation of DNA or RNA by addition of ethanol and salt. Used primarily as a means of concentrating nucleic acids.

Ethidium bromide A fluorescent dye that binds to nucleic acids and enables them to be visualized in UV radiation.

Formaldehyde A chemical that can be used to denature RNA molecules during electrophoresis, to ensure that they migrate according to their true length.

Gel electrophoresis A technique in which charged molecules are separated on the basis of their size, by inducing them to migrate through a gel matrix under the influence of an electric field.

Genome One complete set of genetic material of an organism.

Genomic library A collection of cloned DNA fragments that together represent the entire genome of a particular organism.

Glyoxal A chemical that can be used to denature RNA molecules during electrophoresis, to ensure that they migrate according to their true length.

Grunstein–Hogness screening A technique for detecting colonies of *E. coli* containing recombinant plasmids (or cosmids) that carry a particular insert, by means of hybridization with a labelled probe.

Hybridization Formation of a double-stranded nucleic acid molecule by base pairing between two complementary strands.

Hybridization probe A labelled nucleic acid molecule that can be used to detect complementary nucleic acid sequences by forming stable base paired hybrids.

Manifold A piece of apparatus used in dot/slot blotting.

Nitrocellulose filter A solid support used in blotting procedures to bind nucleic acids. For most applications, it is preferable to use a nylon membrane.

Northern blot A technique for transferring RNA from an agarose gel to a solid support, such as a nylon membrane or nitrocellulose filter.

Nuclear run-on probe A labelled probe synthesized by allowing nuclei to transcribe genes *in vitro*, in the presence of radioactive nucleotides. Used to determine levels of transcription.

Nylon membrane A solid support used in blotting procedures to bind nucleic acids. To be preferred over nitrocellulose filters for most applications.

Plaque A zone of lysed cells in a lawn of *E.coli* resulting from infection by bacteriophage λ.

Plasmid Typically, a circular, double-stranded DNA molecule capable of replicating in *E. coli*. Used as a vehicle, or vector, for cloning DNA fragments.

Polymerase chain reaction (PCR) A technique for amplifying a specific piece of DNA without recourse to cloning.

Positive pressure blotting A technique for performing Southern or northern blotting, in which nucleic acids are forced out of a gel and on to a membrane, by applying increased air pressure above the gel.

Restriction endonuclease An enzyme that recognizes a specific nucleotide sequence in double-stranded DNA and catalyzes the cleavage of the sugar-phosphate backbone of each strand.

Ribonuclease (RNase) An enzyme that degrades RNA.

Slot blot A variation of the dot blot, in which nucleic acid samples are loaded into rectangular wells.

Southern blot A technique for transferring DNA from an agarose gel to a solid support, such as a nylon membrane or nitrocellulose filter.

UV transilluminator A source of ultraviolet radiation used to visualize nucleic acids stained with ethidium bromide.

Vacuum blotting A technique for performing Southern or northern blotting, in which nucleic acids are sucked through a gel and on to a membrane by applying a vacuum beneath the gel.

Vector A DNA molecule, capable of replicating in a host organism, into which DNA from another organism can be inserted to construct a recombinant DNA molecule.

Western blot A technique for transferring protein from an SDS-polyacrylamide gel to a solid support, such as a nylon membrane or nitrocellulose filter.

YAC vector A vector comprising structural elements of a yeast chromosome, used for cloning very large (typically > 1 Mb) genomic DNA fragments.

Index